The Secret of Life

COSMIC RAYS AND RADIATIONS OF LIVING BEINGS

Georges Lakhovsky

TRANSLATED FROM THE FRENCH
by **Mark Clement**

Martino Publishing
Mansfield Centre, CT
2013

Martino Publishing
P.O. Box 373,
Mansfield Centre, CT 06250 USA

ISBN 978-1-61427-507-7

Cover design by T. Matarazzo

Printed in the United States of America On 100% Acid-Free Paper

The Secret of Life

COSMIC RAYS AND RADIATIONS OF LIVING BEINGS

Georges Lakhovsky

TRANSLATED FROM THE FRENCH

by Mark Clement

LONDON

WILLIAM HEINEMANN

(MEDICAL BOOKS) LTD

1939

CORRIGENDUM (page 27)

In the second edition of his major work, Dr. Haviland modified his first statement as follows :

(1) That the districts which had the lowest mortality from cancer were characterised geologically by the older (*a*) (Palæozoic) and most elevated rocks, such as the *Lower* and *Upper Silurian*, and the Carboniferous *Limestone* series ; by the secondary (Mesozoic) *Limestones* of the Oolite (*b*) and chalk formations (*c*).

According to expert opinion, even this statement is not sufficiently explicit and should be further clarified as follows :

(*a*) Older, *i.e.*, primary.
(*b*) Oolitic Limestones of the Jurassic.
(*c*) Chalk formations of the Cretaceous.

PREFACE

By Professor d'Arsonval

What are you thinking about, Faraday ? If I were to tell you, my dear Deville, you might think I was suffering from hallucinations.

Such is the legend.

More confiding than Faraday, Lakhovsky has told me the gist of his ideas on radiations and their effects on living beings. He thought, and rightly, that his ideas could not shock an experimenter who, for the past thirty-five years, had studied the effects of the full range of Hertzian waves on animals and microbes.

In scientific research it is advisable to encourage what appears to be the most daring ideas. I have lived in the intimacy of two great men : Claude Bernard and Brown-Séquard, who revelled in new ideas. And it did not pay them too badly !

The phenomena of resonance have long been familiar to physiologists. We all know the acoustic resonators of the organ of Corti, the optic resonators of the retina since the famous researches of Helmholtz. And still more familiar to us, the *biological resonators* of Charles Henry. Lapicque, Latzareff and myself have invoked on several occasions the phenomena of *cellular resonance* in order to explain the action of nervous influences or other physical agents in living beings.

That space is full of forces which are unknown to us, and that living beings emit radiations or effluvia of which we are not aware, but whose significance has attracted the attention of certain observers, are facts that I have long since accepted. Anything is possible. But one must not accept anything except that which can be experimentally proved. The ideas of an insane person differ from the conceptions of a genius mainly because experiment invalidates the former and confirms the latter.

Lakhovsky, encouraged by his own researches and the practical results he has obtained, is particularly anxious that his theories should rouse interest and stimulate experimental work among independent investigators. Lakhovsky's theories constitute what Claude Bernard called " working hypotheses."

In the " Secret of Life " Lakhovsky confines himself to the study of electromagnetic waves, deeply penetrating waves and unknown waves.

There are certainly many more processes of energy transmission besides those revealed to us by Newton and Fresnel. It is in the study of human beings that the chances of discovering such processes are most promising. Therefore, let us experiment by using the methods of physicists and chemists, and let us aim at discovering the special detector mentioned in the conclusion of this work.

D'ARSONVAL.

CONTENTS

CHAP.		PAGE
	Preface by Professor d'Arsonval . .	v
	Introduction by the Author . . .	1
	Translator's Introduction . . .	5
I.	The Problem of Instinct or Special Sense in Animals	31
II.	Auto-electrification in Living Beings .	43
III.	Universal Nature of Radiation in Living Beings	48
IV.	On Radiations in General and on Electro-magnetic Waves in Particular . .	55
V.	Oscillation and Radiation of Cells .	69
VI.	Modifications in Cells and Oscillatory Disequilibrium	79
VII.	Nature of Radiant Energy . . .	102
VIII.	Sunspots and Cosmic Radiation in Relation to Health and Life	114
IX.	Influence of Nature of Soil on Field of Cosmic Waves—Contribution to the Causation of Cancer—Geological and Geographical Distribution of Cancer—The Rôle of Water in Relation to Cancer	123
X.	Therapeutics of Cellular Oscillation .	140
XI.	Origin of Life	149

PAGE

APPENDIX BY THE TRANSLATOR

I. NOTE ON RADIUM 165

II. THE MULTIPLE WAVE OSCILLATOR . . 176

III. CLINICAL REPORTS 181

IV. EFFECTS OF OSCILLATING CIRCUITS ON ANIMALS 196

INDEX OF NAMES 197

SUBJECT INDEX 199

INTRODUCTION

I SHOULD like to indicate in some way in this introduction the philosophy of my new theory which forms the theme of the present work.

What is the use of propounding a new theory of life? From the beginning of the world have not philosophy and science professed to enlighten us in that respect? What remains of these well-meaning efforts? To the philosopher, and particularly to the metaphysician, I will not attempt to prove the use of a new conception. They know better than I do with what avidity we all welcome the hope of a clearer explanation, the hope of progress in knowledge of the absolute. The craving of the human desire is enough to justify the novelty of a hypothesis. It is the average man, and especially the man of science, that I want to convince. Human knowledge of a positive character is not solely made up, as some would have us believe, of a mass of experimental facts. These facts, by themselves, are worth nothing without the idea which consolidates, arranges and classifies them. The future of science lies essentially, in its dynamic sense, in the expansion of its fundamental concepts, that is to say in scientific hypothesis. Every science is an experimental field whose inter-relations with neighbouring fields, that is to say with other sciences, are more or less unusual and difficult to interpret. Medicine, biology, the natural sciences, are intimately related and their ramifications extend to the domain of chemistry. On the other hand, they seem to be still separated, sometimes by watertight compartments, from the physical sciences, notably from electricity and radio-electricity.

Every progress in the evolution of knowledge shows a new point of view and enables us to explore further the whole field of different sciences, to know their various states of advancement, to observe their mutual relations and the assistance they can render one another.

The most recent discoveries in physics have enabled us to reduce to unity the various phenomena susceptible of analysis through the study of all known radiations. This new field is singularly fertile if one bears in mind that all the most recent discoveries in physics, and consequently in the applied sciences, belong to the domain of radiations : ionic, electronic and atomic ; the usual electromagnetic radiations, radio-electricity, wireless telegraphy and telephony.

Up till now this original conception of radiation, which seems to be the basis of all positive knowledge, has been confined to the realm of the physical sciences and, apart from an incursion into industry, it has not made any important contribution to the natural sciences whose development appears to be limited to that of organic chemistry.

I believe that the time has come to extend the field and the resources of biology by utilising new instruments based on the latest advance of the physical sciences. My theory of the origin of life, which forms the theme of the present work, stands for this new concept uniting two domains of science hitherto kept apart.

Numerous hypotheses, on which I shall not insist, have been advanced to explain the origin of life and various biological phenomena. Let us point out that the most recent of such hypotheses attempt to simplify the problem by reducing these complex phenomena to purely chemical or mechanical phenomena. In view of the unprecedented development of the new discoveries in physics, the latest biological hypotheses appear to be somewhat too simple. Moreover, from the point of view of a higher criterion, they do not give a satisfactory explanation of certain fundamental phenomena which my theory succeeds in doing.

Let us glance at some of those obscure points in biology which we wish to elucidate. Among the most carefully studied facts by naturalists and entomologists, we find all those which are related to the problem of instinct or special sense of animals ; in spite of the accumulation of experimental data, accurate and indisputable, no clear explanation has yet been given of instinct. My theory of radiation of living beings, confirmed by conclusive experiments, is in perfect harmony with the facts in question whose hidden significance is also made clear. Similarly, the rôle of orientation in the flight of birds, the problem of migration, are explicable by the phenomena of auto-electrification in living beings.

What then is this universal radiation in living beings ? My theory expounds in simple terms its fundamental principles and discloses its nature. In deriving support from the most recent discoveries in the domain of radiations, my theory demonstrates, with the aid of elementary analogies, that the cell, essential organic unit in all living beings, is nothing but an electromagnetic resonator, capable of emitting and absorbing radiations of very high frequency.

These fundamental principles cover the whole field of biology.

What is life ? It is the dynamic equilibrium of all cells, the harmony of multiple radiations which react upon one another.

What is disease ? It is the oscillatory disequilibrium of cells, originating from external causes. It is, more especially, the struggle between microbic radiation and cellular radiation. For the microbe, unicellular organism, acts also by virtue of its radiation. If microbic radiation is predominant, disease is the result, and when vital resistance is completely overcome, death occurs. If cellular radiation gains the ascendant, restoration of health follows.

The importance of my theory becomes more apparent in view of the confirmation of its validity as shown by recent experiments on cancerous plants. The recorded cures would seem to give new hope in the treatment of cancer, that terrible disease against which we appear to be struggling in vain. The practical application of my theory, which enables the cells to regain the full vital activity of their radiations, will, in my opinion, give rise to a specific treatment of cancer, in particular, and be equally applicable to diseases due to old age in general.

Apart from its immediate practical applications, my theory may be said to explain, thanks to the rôle played by penetrating radiations, the process of the origin of life, the differentiation of cells and of living species, the phenomenon of heredity, in a word all the great problems whose totality constitutes the biological sciences. I have quite intentionally given a very simple form to the account of my theory, so that it may be understood by all those who have the desire to probe further into the mysteries of science. I have excluded from it any unnecessary phraseology as well as most of the technical terms that cumber the vocabulary of biology and electricity.

The technical terms used in the text of the present work are familiar to all radio listeners. Of such I may single out *self-inductance* which characterises the electromagnetic induction of a circuit ; *capacity* characterising its electrostatic induction ; *electric resistance* which signifies opposition of the circuit to the passage of current ; *wavelength* and *frequency* which characterise the nature of radiation. Mathematical formulæ have likewise been omitted. All relevant scientific explanations are given in footnotes which, however, are not indispensable for understanding the main facts.

My only wish is that my work may be understood by all, even by those who are not familiar with scientific literature. I shall be more than gratified if I have succeeded in my attempt.

GEORGES LAKHOVSKY.

TRANSLATOR'S INTRODUCTION

PAGE

I. PRELIMINARY REMARKS 5

II. GENERAL PRINCIPLES

(a) Definitions 6

(b) Diagrams of Oscillating Circuits . . . 7

(c) Mechanism of Cellular Oscillation . . . 10

(d) Radiations of Living Beings 11

III. COSMIC RADIATION AND SUNSPOTS 13

IV. THE PROBLEM OF CANCER

(a) Lack of Public Support in Cancer Campaign . 17

(b) Limitations of Orthodox Methods of Research . 21

(c) Lakhovsky's Theories on Causation of Cancer; "Multiple Wave Oscillator" in the Treatment of Cancer 24

(d) Cancer in Relation to Soil 25

V. SIMILARITY OF LAKHOVSKY'S AND CRILE'S THEORIES . 28

CONCLUSION 29

I. PRELIMINARY REMARKS

A BOOK, sponsored by Professor d'Arsonval, one of the greatest scientists of our age, must possess exceptional merits and therefore commands attention. On several occasions this world-famous scientist has presented communications to the Académie des Sciences on behalf of Lakhovsky, a talented and independent investigator.

Georges Lakhovsky is a Russian-born engineer who established himself in France, became a naturalised French citizen and was awarded the red ribbon of the Legion of Honour for his technical services during the war.

The "Secret of Life" was originally written in French more than ten years ago. It has been translated into German, Italian and Spanish. An English translation has been overdue for some time and though it is the last in chronological order yet it is the most up-to-date version of all for it contains a great deal of new material which has necessitated the addition of an appendix. The English version is also the only one containing the remarkable photographs of cases treated by means of Lakhovsky's famous apparatus, the Multiple Wave Oscillator.

On the Continent Lakhovsky's work has attracted a great deal of attention in scientific circles, particularly in Germany and in Italy. Italian investigators were among the first to study Lakhovsky's theories and to put them to the test in laboratories and clinics. They were prompt in obtaining a number of significant results.

It was, of course, to be expected that the new science of Radiobiology should make a special appeal to the intellectual heirs of Galvani, Volta and Marconi. It is somewhat depressing to observe that in the country of Faraday and Clerk Maxwell, the theories of Lakhovsky have not yet received the attention they deserve. We may possibly find comfort in the reflection that though the tempo of progress among us in this field of knowledge has been extraordinarily slow, yet in the long run we may bring off achievements transcending the feats of our more alert rivals. At the present time, however, there is no indication of such a trend, and the main purpose of this work is to stimulate interest in experimental research even at the risk of shattering the foundations of established theories which time has often proved to be but working hypotheses.

Lakhovsky's work has not escaped criticism. In accordance with the traditions of orthodox medicine Lakhovsky has been subjected to obstructionism and tyranny on the part of those who invest themselves with the prerogatives of inquisitors. The story of their machinations has been told by Lakhovsky in his work "La Cabale," whose title is sufficiently explicit to leave no doubt as to the nature of its contents. As he himself expressed it: "I have been attacked by physicists ignorant of biology and by biologists ignorant of physics who consequently can neither understand my theories nor judge my experiments."

These self-appointed censors of knowledge soon found that they were confronted with a redoubtable opponent who realised the value of experimental evidence, whose publication made them irate, but not speechless. Lakhovsky retorted by renewed counter-attacks with his oscillating circuits supported by astonishing photographs of regenerated living tissues. This incensed the custodians of infallible doctrines who made up with carping verbiage what they lacked in clarity of vision. But achievements cannot be exploded by verbal fireworks and as evidence accumulates the force of academic petards must needs disperse itself into oblivion. Meanwhile the indisputable fact remains that Lakhovsky was the first to make use of high-frequency electromagnetic waves in the domain of biology. Thus, out of the application of radio-electricity to biology his work developed and gradually established the foundations of the new science of Radiobiology.[1]

* * * * * *

II. GENERAL PRINCIPLES

Physicists and biologists are in the habit of pursuing their researches without encroaching upon one another's domains. As an independent investigator Lakhovsky was not confined to any particular field, and having formulated his theory of cellular oscillation, the fusion of physics and biology was the natural consequence.

Lakhovsky's original experiments on cancerous plants are of great scientific importance and constitute a landmark in the history of radio-electrical methods of treatment. This has been fully acknowledged by leading electro-therapists on the Continent and in this country. In his excellent work on Diathermy,[2] Cumberbatch states: " Although it had frequently been observed that the short Hertzian waves could produce heat at a distance from the transmitter, the first scientific investigation on the subject from a biological point of view was made by Gosset, Gutmann and Lakhovsky. In 1924 they published a paper on the effects of very short

[1] The First International Congress of Electro-Radiobiology was held in Venice in 1934.

[2] " Diathermy," E. P. Cumberbatch, M.A., M.B., F.R.C.P. (Heinemann (Medical Books) Ltd., 3rd edition, 1937.)

waves on cancer in plants. The wavelength was about two metres, corresponding to a frequency of oscillation of 150,000,000 per second."

* * * * * *

(*a*) Before considering the subject of cellular oscillation, which forms the basis of Lakhovsky's work, it is of the utmost importance that the reader should understand the principles underlying the mechanism of an oscillating circuit. Indeed, Lakhovsky's "Secret of Life" presupposes a certain amount of knowledge of electricity, wireless phenomena and biology. Nowadays most people are familiar with technical terms used by radio engineers, such as Hertzian waves, frequency, transmitter, inductance, triode valve, etc. These terms are not jargon; they do not make science more complicated for the average man; on the contrary they simplify it by introducing order and precision. But a certain mental effort is required to understand what these terms really denote, and for the sake of those who have little biology and less physics, and who are yet desirous of knowing something about wireless phenomena, cosmic rays and radiations emitted by living beings, a clear explanation of the standard technical terms used in this work is imperative, or else the "Secret of Life" will not be revealed to them.

The technical term most frequently used in the course of this work is *oscillating circuit*. It occurs again and again throughout the text and therefore demands special attention. A formal definition of this term is thus required at the outset.

An oscillating circuit is a circuit containing *inductance* and *capacity*, which, when supplied with energy from an external source, is set in electrical vibration and oscillates at its *natural frequency*.

As the terms inductance and capacity are not self-explanatory they must be further defined.

A conductor is said to possess *inductance* if a current flowing through it causes a *magnetic field* to be set up round it. A straight wire has inductance; if a conductor is wound up in the form of a coil the value will be greatly increased.

The *capacity* of a condenser or isolated body is a measure of the charge or quantity of electricity it is capable of storing.

(*b*) A diagram of an oscillating circuit will now make things clearer. C_1 and C_2 represent the plates of a condenser placed in a circuit containing an inductance coil I. There is a gap in the circuit at S.G. known as the spark-gap.

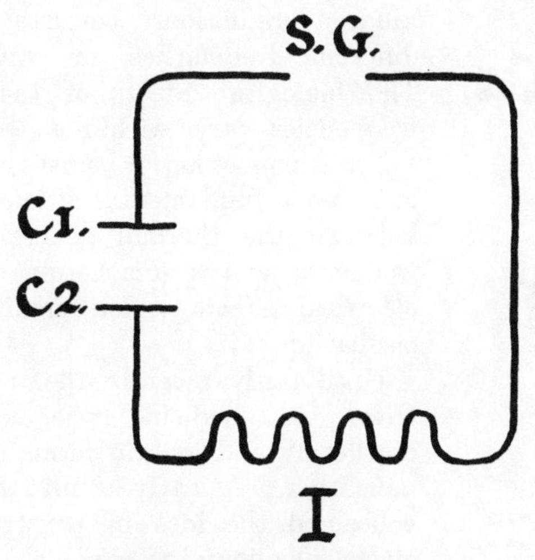

Diagram of Oscillating Circuit.

Now when the condenser is charged and the difference of potential between the two plates is sufficiently high, a spark will appear at the gap and the condenser will be discharged. While the spark appears at the gap a current will oscillate in the circuit.

The inductance coil I is called the oscillator, and the circuit containing the condenser, oscillator and spark gap is called the *oscillating circuit*.

If the condenser plates are opened out so that the induction coil lies between them, we have an " open " oscillating circuit.

From such a circuit energy is readily given off in the form of waves. By taking a suitable condenser and inductance coil, the frequency of oscillation may be raised to any required value.

*   *   *   *   *   *

(*c*) We are now in a position to discuss the subject of cellular oscillation on which Lakhovsky's theories are based.

From a physical point of view the essential difference between the various kinds of electrical oscillations and kindred radiations consists in their different frequencies or wavelengths. The biological effects of the different frequencies vary within a wide range. In this connexion it must be pointed out that a fundamental difference exists between the thermal effects of high-frequency waves (diathermy) and their electrical effects which modify cellular oscillation.

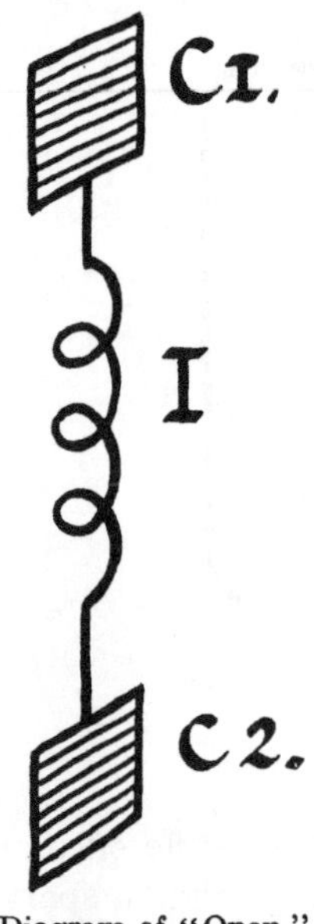

Diagram of "Open" Oscillating Circuit.

Until fairly recently the use of electricity in medicine was confined to diathermy and certain forms of darsonvalisation. As early as 1923 Lakhovsky conceived the idea of constructing an electrical apparatus capable of emitting continuous waves of very short length (2 to 10 metres). His intention was to demonstrate that the living cell was like a wireless apparatus, having the dual power of transmitting and receiving waves. This apparatus, known as the Radio-cellulo-oscillator, was first used for the treatment of experimental cancer in plants at the Surgical Clinic of the Salpêtrière in Paris. The results were so conclusive that a special communication was addressed to a leading scientific society drawing attention to the remarkable cures obtained by this new mode of electrical treatment.

The Radio-cellulo-oscillator was the prototype of various electrical generators of short waves that finally led to the invention of the Multiple Wave Oscillator, fully described in the Appendix.

* * * * * *

The fundamental principle of Lakhovsky's scientific system may be summed up in the axiom "Every living being emits radiations." Inspired by this principle Lakhovsky has been able to explain such diverse phenomena as instinct in animals, migration of birds, health, disease, and in general, all the manifestations of organic life.

According to Lakhovsky, the nucleus of a living cell may be compared to an electrical oscillating circuit. This nucleus consists of tubular filaments, chromosomes and mitochondria, made up of insulating material and filled with a conducting fluid containing all the mineral salts found in sea-water. These filaments are thus comparable to oscillating circuits endowed with capacity and self-inductance and therefore capable of oscillating according to a specific frequency.

In the light of Lakhovsky's theories, the fight between the living organism and microbes is fundamentally a "war of radiations." If radiations of the microbe win, the cell ceases to oscillate and death is the ultimate result. If, on the other hand, radiations of the cell gain the ascendant, the microbe is killed and health is preserved. Broadly speaking, health is equivalent to oscillatory equilibrium of living cells while disease is characterised by oscillatory disequilibrium. This general principle has given rise to a vast number of experiments covering the whole field of biology.

* * * * * *

(*d*) During recent years observations on the part of several investigators appear to have established the fact that most animals, including insects and birds, emit radiations while they are also sensitive to the influence of external electromagnetic waves. These radiations, emitted by living beings, cover an indefinite range and are characterised by a multitude of different wavelengths. The luminescence of the glow-worm is an example of vital radiation perceptible to our visual sense. In the immense range of existing vibrations we can only perceive the luminous octave, but we know that whole gamuts of radiations exist beyond the narrow limits of the visible spectrum. In the face of a mass of cumulative evidence it is perfectly rational to conclude, as Lakhovsky does, that the phenomenon of radiation is a universal property of living matter even as radioactivity appears to be a universal property of inanimate matter. The limitations of our senses prevent us from perceiving radiations of living beings while this sensory incapacity also excludes from the field of direct awareness the vast gamut of electromagnetic waves traversing our atmosphere. Yet all these radiations and waves exist and affect all forms of life in various ways.

These views can no longer be regarded as speculative, for radiations of living beings have actually been apprehended by means of the spectroscope and the photographic plate. The *mitogenetic* radiation given off by rootlets of growing plants and vegetables has been identified by Gurwitsch and Frank as belonging to the ultra-violet region of the spectrum. The original experiments of these two investigators established the fact that the stimulus to growth was oscillatory in character and that it was associated with a specific wavelength, thus confirming the fundamental principles of Lakhovsky's theories. Further experiments by Reiter and Gabor have shown that embryonic tissues and malignant tumours possessed a high radiation potential varying in intensity according to the rate of growth. These experimenters succeeded in measuring the wavelength of radiating tissues and in modifying the development of selected organisms by subjecting them to the influence of a certain range of ultra-violet rays.

As all these vital radiations appeared to be akin to ultra-violet rays in regard to frequency, the problem of photographing them came under consideration. In spite of great technical difficulties, Professor Guido Cremonese succeeded in making photographic records of radiations given off by living substances which were duly reproduced in a valuable monograph.[1] With regard to human radiations,

[1] Guido Cremonese, I Raggi della vita fotografati. Rome, 1930.

Professor Cremonese devised means of making direct photographic records, or failing that, he made use of specimens of saliva or blood which were shown to be active sources of radiations.

Professor Cremonese's investigations have opened up a new field of scientific research whose future developments may have a most important bearing on the problems of biology and medicine.

* * * * * *

III. COSMIC RADIATION

The subject of cosmic radiation is given such a prominent place in Lakhovsky's theories that it calls for a few introductory remarks. Only the briefest outline of this aspect of cosmography can be given here, but its importance in the universal scheme in relation to living processes should always be borne in mind.

The study of cosmic rays is of recent origin, the first significant observations having been published in 1900. Since then this subject has attracted a host of famous investigators in Europe and in America.

In the study of cosmic radiation American physicists have played a pre-eminent part. It is largely due to the labours of Professor Millikan and his associates that the knowledge of cosmic rays has made such rapid progress during the past decade.

Cosmic rays is the name given to a highly penetrating radiation travelling through our atmosphere and emanating from a far distant source. This radiation is much more penetrating than X-rays and is correspondingly shorter in wavelength. According to Millikan, cosmic rays are of the nature of electromagnetic radiations similar to light but of extremely short wavelengths. He has suggested that they might be the result of aggregations of Hydrogen atoms into Helium, a process constantly going on in the universe. In Millikan's picturesque phrase, cosmic rays are the "birth cries of atoms." He estimated that the "total radiant energy in the universe existing in the form of cosmic rays is from 30 to 300 times greater than that existing in all other forms of radiant energy combined."

The amount of this radiation must obviously be very great. Sir James Jeans[1] states that "it must break up millions of atoms in each of our bodies every second—and we do not know what its physiological effects may be." In a recent discussion at the Royal Society of Medicine,[2] it was admitted that not enough notice had been taken of electrical conditions of the atmosphere as affecting the human organism. Investigations had been made on the effect of cosmic radiations in influencing the mutation of insect species and the available experimental evidence furnished a basis for correlating the periodicity of epidemics with cosmic radiations.

The consensus of opinion among astrophysicists favours the view that cosmic rays are high-speed charged particles (electrons, protons, positrons, α-particles, etc.), and it is thought that these particles are associated with some kind of radiation of extremely high frequency.

Cosmic radiation is the most penetrating form of radiation known to us. Many expeditions have been organised to study the geographical distribution of cosmic rays which appears to vary according to latitude and altitude.

The relative penetrating powers of different cosmic rays were ascertained by sending automatic recording electroscopes up to 29,000 feet in airplanes and up to 60,000 feet in balloons, while under-water experiments showed traces of the most penetrating cosmic rays at depths as low as 770 feet (Regener, 1931).

The foremost investigator of cosmic rays in this country, Professor Blackett, has devised an apparatus whereby cosmic rays take photographs of themselves or of their own tracks. The researches of Blackett and Occhialini have definitely confirmed the existence of the positive electron or positron, originally discovered by Anderson.

The particular importance of cosmic rays in science, apart from their very high energy, is their practical use for the investigation of atomic structure.

According to Lakhovsky, the geological nature of the

[1] Sir James Jeans, "The Universe around Us," 3rd edition. University Press, Cambridge, 1933.

[2] *British Medical Journal*, March 7th, 1936.

soil modifies the field of cosmic radiation at the earth's surface and this gives rise to secondary radiations which must be taken into account in biological phenomena. It would appear that cosmic radiations have a direct effect on all living cells, their function being to maintain, by resonance and interference, the natural oscillation of healthy cells, and to re-establish the vibrations of diseased cells by neutralising antagonistic radiations such as those given off by microbes. Moreover, an excessive amount of cosmic radiation may prove detrimental to living organisms. In order to remedy this contingency Lakhovsky devised a special type of oscillating circuit which, by creating an auxiliary electromagnetic field, acts as a " filter " of cosmic waves and protects the organism against the harmful effects resulting from excessive radiation.

Variations in the field of cosmic radiation bring about a state of disequilibrium in living cells which can no longer oscillate according to their natural frequency. The ultimate result of this sequence of events may manifest itself in the distressing phenomenon of cancer.

* * * * * *

SUNSPOTS

Galileo was the first to study sunspots in a scientific manner, and by measuring their motion he proved that the sun was rotating.

In one of the most fascinating works of Sir James Jeans [1] we are told that " sunspots are of the nature of vent holes from which masses of hot gas are shot out at terrific speeds. The matter which they eject is probably a mixture of complete atoms and fragments of atoms which may include electrified particles of various kinds. They are shot out and travel in all directions, some of them will reach the earth, and, penetrating its atmosphere, may produce a display of the Aurora Borealis. Later they may ionise the air and so form the layers which reflect our wireless waves back earthward and enable us to hear distant wireless stations. . . . Sunspots do not come in a steady stream but

[1] Sir James Jeans, " Through Space and Time."

rather in gusts or waves, their numbers fluctuating up and down every eleven years or so. Sunspots were especially numerous in 1906, 1917 and 1928 and will be so again in 1939."

At the beginning of the nineteenth century Sir William Herschel noted that during the period 1650 to 1713 a scarcity of vegetation, judging by the normal wheat yield, had occurred whenever the activity of sunspots was lowest.

Lakhovsky looks upon sunspots as an important source of cosmic radiation. He has shown that the curves representing activity of sunspots, frequency of magnetic perturbations and aurora borealis, are strikingly parallel. He has also established a correlation between sunspots and good vintage years which appear to synchronise with maximal solar activity.

Some years ago Dr. Meldrum, Director of the Observatory at Mauritius, established the fact that the number of cyclones in the Indian Ocean and the West Indies varied with the sunspot area. It also appeared that the occurrence of maximum rainfall coincided with the occurrence of maximum sunspots. This relation was so definite that it could not be regarded as accidental. These observations were confirmed by Sir Norman Lockyer, who realised that not only the rainfall but also many other features of the earth's physical condition were connected with the sunspot cycle.

Since then astrophysicists have correlated the frequency and intensity of sunspots with a certain number of physical phenomena, while recent research has extended this correlation to biological phenomena, particularly those pertaining to human pathology.

Colonel C. A. Gill [1] recently pointed out that pandemics of malaria from 1800 onwards had occurred at a period of minimal sunspot activity. Every pandemic of malaria since sunspot records were taken had occurred at a time when sunspot numbers were lowest. The same phenomenon had been observed in connexion with yellow fever; the epidemics in East Africa since 1800 had also occurred at a time of minimum sunspot numbers.

[1] *British Medical Journal*, March 7th, 1936.

Further evidence is given by Dr. Conyers Morrell,[1] who states that waves of epidemic disease covering considerable periods exhibit a very close correspondence with the phases of sunspot periods. He points out that close correlations over long periods have been found to occur in several diseases, notably diphtheria, typhus and dysentery in Russia and Denmark, and plague in India.

Tchijevsky has shown a close correspondence between mortality from diseases of the nervous system and the sunspot curve, while an equally close connexion between the frequency of epileptic fits and the incidence of solar storms has been reported by other investigators.

Many phenomena such as abundance of harvests, growth of trees, migration of birds, lake levels, blizzards and radio reception show a well-marked relationship with meteorological periodicities conditioned by the solar cycle which may well play an important rôle in influencing health and disease in human beings.

Dr. Conyers Morrell deplores the fact that " while in other countries considerable advance has already been made in the institution of enquiries into the causal relationship between solar activity and terrestrial phenomena, in this country this very important and interesting subject has received as yet but very scant attention, and has, in fact, like other revolutionary but now accepted hypotheses, met with a marked quantum of the scepticism so inseparable from the conservatism of English scientific enquiry."

We shall see in the next chapter dealing with the cancer problem, to what extent this conservatism obscures vision and cripples research to the detriment of the whole community.

* * * * * *

IV. THE PROBLEM OF CANCER

(*a*) In the " Secret of Life " the subject of cancer occurs at frequent intervals like the *Leitmotiv* in a musical composition. The theme is developed from a biophysical point of view and the malignant cell becomes a centre of activity

[1] *British Medical Journal*, March 14th, 1936.

where electromagnetic waves, cosmic rays and cellular radiations play a predominant part. For many years Lakhovsky has concentrated his attention on cancer and has evolved original theories which should be judged by the practical results he has obtained. As cancer is the main objective in Lakhovsky's work, it demands a correspondingly extensive treatment on our part so that the reader may realise the full significance and the complexity of the factors involved in the causation of that dreaded disease.

Many cancer research workers, including the present writer, have found that the vast majority of people show a profound aversion to discussing the subject of cancer and treat it *as if* it did not exist. This general flight from reality renders the solution of the cancer problem even more difficult than it actually is, for fear and ignorance on the part of the public are not in the least helpful to those who strive to save their fellow-beings from the tentacles of cancer.

A rational and constructive attitude is incumbent upon everyone. Cancer must be faced as a stark reality, a constant threat to all, ignored only at the peril of our own lives.

If the problem of cancer is ever to be solved, the public, that is to say, everyone of us, must take an intelligent interest in it and not rely on Providence to do our own salvaging work.

Cancer is a plague as deadly as any plague that has ravaged mankind in the past. Indeed, its insidious onset and absence of symptoms in the early stages make it a far more treacherous disease than tuberculosis and less responsive to treatment, since it is often far advanced before the victim becomes aware of his hopeless condition.

It is of the utmost importance that the public should realise the magnitude of the cancer problem, for the medical profession is in urgent need of psychological as well as of financial co-operation in this respect, neither of which has been forthcoming in anything like an adequate measure. A leading London surgeon, and one of the foremost cancer investigators in this country, has put it on record that " the sum yearly available for cancer research does not amount to one penny per annum for each individual composing the population of Great Britain." [1] This contribution becomes still more preposterous when it is realised that one person in six, over the age of 40, is doomed to die of cancer. When we consider the colossal sums of money spent on protecting ourselves against the possibility of foreign aggression, it seems incredible that so little is given to fight against cancer, the ruthless enemy in our midst, inflicting a death-roll that goes on increasing without respite. It is clear that the policy of voluntary contributions for the maintenance of hospitals and associated services is now obsolete and that its survival is nothing but a sentimental tradition incompatible with actual exigencies. At last there are signs of realisation that adequate financial assistance on the part of the authorities is an imperative necessity in order to deal with the social problem of cancer which decimates the community and causes incalculable suffering. In this connexion a glance at official statistics will show the actual state of affairs.

The certified deaths from cancer for England and Wales during the quinquennium 1932–1936 were as follows: [2]—

1932	60,716
1933	61,572
1934	63,263
1935	64,507
1936	66,354
	316,412 [3]

In addition to this soaring mortality it should be borne in mind that for every death from cancer there are at least three living cases of the disease. Thus, at any given time, there are in our midst 200,000 people of all ages suffering from cancer.

We are not here concerned with the ever-recurring question whether cancer is on the increase, nor are we particularly impressed by the official statement that the

[1] "Cancer Research at the Middlesex Hospital (1900–1924)."
[2] Annual Reports of the Chief Medical Officer of the Ministry of Health.
[3] The latest figures for Great Britain (1937) show a total cancer mortality of 74,000. (*Times*, December 13th, 1938.)

increase is more apparent than real, a dialectical balm dispensed by bureaucrats to allay fears while evading main issues. Among those best qualified to express an authoritative opinion, the late Lord Moynihan, President of the Royal College of Surgeons, called attention to the fact that statistics will prove anything, even the truth ! Lord Moynihan stated "There is little doubt that cancer is definitely on the increase especially in certain organs. In the last seventy years the mortality from cancer has increased fivefold." [1] The vital fact confronting us now is that in the last five years over a quarter of a million people have died of cancer in this country and that they go on dying at the rate of nearly 70,000 per annum. This means that *every eight minutes, day after day throughout the year, someone dies of cancer in England.*

Such a disastrous toll of life calls for an intensive campaign, organised with sustained efficiency, in which the public must co-operate to the limits of their powers. At the present time the Government is contemplating a Bill for the establishment of a National Cancer Service, making available modern facilities for diagnosis and treatment over the whole of England and Wales. This belated measure must be acclaimed as a practical contribution to the cancer campaign, but it cannot possibly succeed in checking the overwhelming stream of new cases of cancer which constitutes the crux of the problem.

How this is to be achieved is a matter beyond the scope of this introduction, the writer having elaborated an appropriate policy in another work. Suffice it to say that research is seriously hampered by financial considerations and by adherence to laboratory ideologies as outworn as they are barren. Though pyramids of data are unceasingly erected by a host of servants of the dominant methodology, yet their labour, in the words of Bacon, has been rather in circle than in progression. Its effect on reducing the cancer death-rate remains negligible.

The solution of the cancer problem in its manifold aspects demands a mood of stern realism unsullied by any trace

[1] Addresses on Surgical Subjects by Sir Berkeley Moynihan, Bart. London, 1928.

of professional complacency. Too much stress is laid on treatment while methods of preventing cancer remain pious aspirations. So long as this attitude predominates there can be no hope of achieving the conquest of cancer. The example of tuberculosis should serve as a timely reminder that the mortality of the " white plague " has been halved within the past fifty years by preventive hygienic measures and not by specific remedies. Moreover, the system of institutional research, like that of any institutional organisation, is threatened by the perils of authoritarianism and ultimate sterility.

The supreme need in cancer research at the present time is the multiplication of individual initiative, unfettered by the cramping influence of official domination, and remunerated at least as adequately as the services of Civil Servants.[1]

The history of Medicine is a long record of discoveries and advances made by men who worked on independent lines and who emancipated themselves from the shackles of conventional doctrines. This is a truth that the public seldom realise and committees always ignore.

Science, like art, is not the outcome of corporate movements, but the result of individual travail in the midst of collective apathy.

* * * * * *

(*b*) The present position of the cancer problem has been admirably summed up by an eminent British surgeon who has spent thirty years studying this question.[2] " Present methods of treatment are purely local and ignore the deadly fact that though they may at best account for some 30 per cent. of cancers, such triumphs of modern science are often accomplished by means of nerve-shattering operations while

[1] We have seen that the annual contribution of the British public towards the fight against cancer was assessed at less than one penny per head of the population. In the course of a recent B.B.C. discussion, we were informed that every season this same public willingly spent £50,000,000 on football pools !

This contrast in expenditure is a striking example of the false values that vitiate our national life and permit every individual to enjoy the freedom of ignorance.

[2] " Tumours and Cancers," Hastings Gilford, F.R.C.S. London, 1925.

cancer in general continues to flourish. In spite of further improvements in operative technique, we are beginning to see that surgery has its limitations and that we are coming to a point whence we can no longer advance."

This indictment is reinforced by a statement made by the Vice-President of the Royal College of Surgeons who said that: "In a series of 5,500 cases of cancer the total survivorship from all cases was 30 per cent. . . . less than 1 in 3. That is all our vaunted surgery can do for our patients! If we can only save 30 per cent. of our cancer patients, something more ought to be done. *There is room for both individual and organised effort.* Surgery, advanced as it is, cannot be expected to do more than it has done. The end results of radium treatment are not much better than those of surgery." [1] (See Note on Radium in Appendix.)

This frank admission of failure on the part of a leading member of the medical profession clears the way for further comments by independent investigators who are profoundly dissatisfied with the prevailing methods of cancer research.

Hastings Gilford, formerly Hunterian Professor at the Royal College of Surgeons, and one of the most learned writers on the subject of cancer, expressed his views with characteristic vigour: "That the research into the cause and nature of cancer is making no headway is obvious to everyone who has followed its drift since the movement began . . . all it has to show is a prodigious heap of facts . . . useless to man. . . . Laboratory cancer research has gone on for so many years contentedly grinding out data and spinning inductions without attention being drawn to the fact that it never produces any useful results. And now, after a quarter of a century of research, we can see to what a deplorable waste of energy, ability and money this academic aimless toil may lead. One useful, if negative induction emerges, which is that the problem of causation of human cancer is not to be solved by experiments in lower animals in laboratories." [2]

[1] *Proceedings of Royal Society of Medicine*, vol. 27, part 2; 1934, Section of Surgery, p. 69.
[2] *Lancet*, October 25th, 1930.

In view of the appalling cancer mortality it behoves us to enquire whether the prevalent methods of cancer research are likely to result in curative and preventive measures that will reduce the incidence of the disease to an appreciable degree, for that is the most urgent desideratum.

It is important to observe that in this particular sphere of scientific activity, namely cancer research, where a strictly objective attitude of mind should be enforced, all kinds of subjective considerations are allowed to gain dominance, thus sidetracking main issues and actually impeding progress. This deplorable tendency has been indicted by John Cope in his stimulating work on "Cancer, Civilisation and Degeneration," recently published. He states that: "Experimental cancer research has become so isolated and so entrenched that, without being aware of it, the researcher now almost instinctively regards those who criticise his opinion, question his authority or adopt other methods of working, not as fellow-workers, but as amateurs, as 'outsiders,' or even as positive enemies. The suggestion that the causes and nature of human cancer are far more likely to be revealed by a study of the habits and customs of human beings in all parts of the earth comes almost as a shock."

It need hardly be added that Lakhovsky, as an independent investigator, was duly treated as an outsider by the medical fraternity in Paris, who, in regard to obstructionism, have little to learn from any professional organisation in the world. The classical example of Pasteur's conflict with the Paris Faculty of Medicine is still vivid in our memories.

Thus it is clear, as the Vice-President of the Royal College of Surgeons boldly affirmed it, that individual efforts are worthy of encouragement in view of the failure of institutional research.

Sir Ronald Ross, whose great work on malaria was carried out on strictly individualistic lines, expressed his own convictions on the subject of research in a very decisive manner. "I believe that institutional research has never yet solved one of the great problems of nature, including those of medicine. *I will venture to predict that it never will solve any of the problems connected with cancer.* The discoverer, like the poet, is an individualist. He must not be controlled by any committee and he must choose his own time and place."

It is highly significant that Lakhovsky achieved his remarkable results in the field of cancer research, working alone in his laboratory and ignored when not attacked by exponents of orthodox cults.

Lakhovsky has repeatedly emphasised the necessity of approaching the cancer problem without preconceived ideas or bias of any kind. As an engineer-physicist he is guided only by experimental evidence and not by medical dogmas masquerading as certitudes in a domain of doubts.

Official medicine is so deeply indebted to physicists for its own progress in the past that it can hardly afford to disregard new theories on the constitution of living cells such as those formulated by Lakhovsky in his " Secret of Life." Both X-rays and radium, without which modern medicine is inconceivable, are the gifts of physicists who were not primarily concerned with the problems of disease. And yet Röntgen and Madame Curie have done more for medical science than a host of academic professors of the healing art.

* * * * * *

(c) Many hypotheses have been advanced to explain the formation of cancerous tumours. Heredity, contagion, local irritation, traumatism, etc. Having formulated his theory of cellular oscillation, Lakhovsky applied it to elucidate the question of cancerous tumours which he attributed to oscillatory disequilibrium. His first experiments bearing on this subject were carried out on cancerous plants as far back as 1924.

According to Lakhovsky the essential cause of cancer formation is to be sought in the oscillatory disequilibrium of bodily cells. But the problem of attempting to establish the equilibrium of the cells composing the human body would appear to be insoluble. Our bodies contain approximately two hundred quintillion cells and in this fabulous number there are not two cells vibrating with the same frequency, this being due partly to the incessant activity taking place within the cells and partly to the specific characteristics of different tissues, not to mention many other factors. Moreover, from a biological point of view, it would be impossible to find at any given time two individual cells exactly alike in every respect. Every cell of every individual tissue of any particular species is characterised by its own oscillation. In order to produce, artificially, an oscillatory shock in disequilibrated cells it would be necessary to generate as many wavelengths as there are different cells in any given body. The problem would thus seem to be insoluble. With remarkable imaginative insight, Lakhovsky finally evolved a solution. To that end he designed a new type of radio-electrical apparatus, his famous Multiple Wave Oscillator, generating a field in which every cell could find its own frequency and vibrate in resonance. The practical results he obtained in various hospitals soon confirmed his theoretical views. Clinical reports and photographs of cases treated with the Multiple Wave Oscillator are included in the Appendix, to which the reader is referred.

Lakhovsky does not claim that his Multiple Wave Oscillator will cure all cases of cancer, for that is obviously impossible by any method of treatment, including surgery and radium. He claims, however, that a number of cases have been definitely cured and that a marked improvement in the general condition of all patients treated with the Multiple Wave Oscillator has frequently been observed even in the most advanced cases. Furthermore, alleviation of the distressing symptoms associated with such cases is by no means a minor achievement.

CANCER IN RELATION TO SOIL

Lakhovsky postulates that in cancerous patients a certain degree of oscillatory disequilibrium in the cells must occur, and this would account for the abnormal development characteristic of neoplastic tissues. Lakhovsky also sought to determine the underlying causes of cancer formation in the light of his theory of cellular oscillation. Among the possible causes he stressed the rôle of external radiations of all kinds susceptible of affecting the human organism, more especially cosmic radiations which, in striking solid bodies, give rise to secondary radiations of longer wavelengths and variable intensity.

In studying the geographical distribution of cancer from official statistics, Lakhovsky was able to establish the fact that the density of cancer incidence was closely connected with the geological nature of the soil. He has shown the relationship between cause and effect and the part played by cosmic radiation whose field at the earth's surface is modified by the nature of the soil according as the latter is an insulator or a conductor of electricity.

Lakhovsky finally came to the conclusion, confirmed by other workers, that soils that are specially permeable to rays, that is to say " dielectric," such as sand, sandstone, gravel, etc., absorb external radiations to a great depth without giving any reaction on the surface, while soils that are impermeable to rays, that is to say conductors of electricity, such as clay, marl, alluvial deposits, carboniferous strata, mineral ores, etc., are resistant to penetration by rays and give rise to secondary radiations which modify the field of external radiations. It is these impermeable soils which are associated with the highest incidence of cancer.

Lakhovsky is quite aware that the causation of cancer is an extremely complex problem and that his theories deal only with certain fundamental aspects of it.

In the development of cancer the human " terrain " is of greater importance than the geological nature of the soil, but that is not to say that the latter may be dismissed as being altogether insignificant, a tendency only too prevalent among laboratory workers who specialise in any particular line of research.

It may be pointed out that Lakhovsky is by no means an isolated exponent of the geological factor in the causation of cancer. Several German investigators have made a special study of this question, notably Behla, Kolb and von Pohl, and their systematic observations still await translation. The only English study at all comparable with that of the German workers is that of Dr. Alfred Haviland, published as far back as 1875 and re-issued in 1892 under the title of " The Geographical Distribution of Disease in Great Britain."

The main facts which Haviland brought to the notice of the medical profession were summarised by him as follows :

(1) That the districts which had the *lowest mortality* from cancer were characterised geologically by the older and most elevated rocks, such as the Lower and Upper Silurian and the carboniferous limestones of the oolite and chalk formation.

(2) That the districts which had the *highest mortality* were characterised geologically by clays, such as the London clay of the Eocene, the Boulder clay of the Pleistocene or Glacial period and the brick earths and alluvial deposits of recent origin.

The high mortality districts were found to be traversed by fully formed rivers that seasonally flooded their banks. Haviland added that " this coincidence of *low mortality* from cancer in localities characterised by *limestones* would not perhaps have been so much dwelt upon had not the same fact occurred throughout England and Wales wherever limestones occur, whilst on the other hand wherever *clays* and *floods* are associated, *high mortality from cancer at once prevails.*"

At the request of the Editor of the *Lancet*, Dr. Haviland wrote a series of articles on the geographical distribution of cancerous diseases in the British Isles, which appeared in that journal in the course of 1888 and 1889.

He concluded that " those who have reason to dread cancer should live in high dry districts characterised by either limestone or chalk formations."

The geological nature of the soil is clearly not the only factor in the causation of cancer. Civilisation has brought in its train many habits and customs whose final results are manifested in physical degeneration forming a suitable terrain for the development of cancer.

People tainted with alcoholism, tuberculosis or syphilis, not to mention the toxic effects of excessive smoking and adulterated foods, constitute a degenerate stock whose cellular oscillations, to use Lakhovsky's expression, are not in a state of equilibrium. This oscillatory anarchy in the human organism has far-reaching repercussions in every organ and living tissue, ultimately causing, under certain conditions, the fatal proliferation of cells characteristic of cancer.

* * * * * *

V. SIMILARITY OF LAKHOVSKY'S AND CRILE'S THEORIES

The theories of Lakhovsky bear a striking similarity to those of Dr. George Crile, the eminent American surgeon, whose great work on surgical shock has earned him an international reputation. In his admirable book entitled "The Phenomena of Life,"[1] Dr. Crile points out that electrical energy plays a fundamental part in the organisation, growth and function of protoplasm. Lakhovsky and Crile, pursuing their investigations independently, have come to identical conclusions. While the engineer-physicist was experimenting with his oscillating circuits, the surgeon was testing in the clinic the principles of radio-electricity.

The foundations of Lakhovsky's theories rest on the principle that life is created by radiation and maintained by radiation. Crile states that man is a radio-electrical mechanism and stresses the significant fact that when life ends, radiation ends. He writes : " It is clear that radiation produces the electric current which operates adaptively the organism as a whole, producing memory, reason, imagination, emotion, the special senses, secretions, muscular action, the response to infection, normal growth, *and the growth of benign tumours and cancers*, all of which are governed adaptively by the electric charges that are generated by the short wave or ionising radiation in protoplasm."

Like Lakhovsky, Crile holds that living cells are electric cells functioning as a system of generators, inductance lines and insulators, and that the rôle played by radiation and electricity in living processes is no more mysterious in man than in batteries and dynamos.

In formulating new concepts it is often necessary to coin

[1] "The Phenomena of Life." A radio-electric interpretation by George Crile. Heinemann. Printed in U.S.A., 1936.

new terms and Crile has used the term *radiogen* to denote the theoretical units of protoplasm in which oxidation occurs and from which radiation is emitted. Crile's radiogens correspond fundamentally to Lakhovsky's *biomagnomobile* units that characterise all living organisms.

With an impressive array of experimental data Crile makes it clear that all activities of the living organism, including those of the brain, nerves, muscles and glands, are dependent upon the specific radiant properties of the visible spectrum. According to this brilliant investigator, ultra-violet radiation plays the most prominent rôle in the body, for it has the greatest power of generating electricity and of ionising atoms essential for building up organic compounds which constitute cellular protoplasm. Furthermore, the control of the special senses is effected through environmental energy, mainly solar radiation. In fact, the entire energy system of living beings is controlled by radiant and electrical forces in the environment. Thus it becomes evident that the "spectrum of the living" reflects innumerable environmental changes and is itself changing continually in consciousness, in sleep, in emotion and in every adaptive reaction.

Both Crile and Lakhovsky have been remarkably successful in the practical application of their theories, which have extended our knowledge of biological processes and revealed a new vista of scientific exploration that may ultimately lead to the solution of the cancer problem.

* * * * * *

CONCLUSION

We began by considering the elementary laws of electricity which we found to be applicable to all living things. The study of radiations of the individual cell led us to contemplate analogous phenomena throughout the universe. This cosmic harmony of vibrations has given birth to the concept of *Universion* which Lakhovsky defines as the synthesis of the infinitely great and the infinitely small. *Universion* consists of the entire plexus of cosmic radiations emanating from interplanetary space. Its nature is indestructible and all-pervading. It is the ultimate reservoir of all matter undergoing the cyclical phases of destruction and reconstruction.

The validity of such speculations may be claimed to have been established by Lakhovsky's experiments extending over the whole range of organic life. The practical application of radio-electrical theories has had the most remarkable results. Cancerous plants have been cured, human tissues regenerated and animals restored to vigorous health. Such are the outstanding achievements due to a solitary researcher working in the face of formidable handicaps aggravated by the opposition of witless reactionaries.

Lakhovsky has accomplished his task and his brilliant work deserves the highest praise. His life and his fortune have been generously devoted to research, but he claims no reward for himself. He merely ventures to hope that his work has widened our intellectual horizon and that it will benefit suffering humanity. Both ends have already been attained in an appreciable measure and there is every reason to anticipate that Lakhovsky's theories will give rise to still more significant developments in the future.

THE SECRET OF LIFE

CHAPTER I

THE PROBLEM OF INSTINCT OR SPECIAL SENSE IN ANIMALS

General Considerations—Instinct of Orientation—Carrier Pigeons—Nocturnal Birds—Bats—Lemmings—Functions of Semi-circular Canals and of Antennæ in Insects—Nocturnal Experiments with the Great Peacock-Butterfly—Diurnal Experiments with the Oak Bombyx—Activities of Burying Beetles (*Necrophorus*).

General Considerations

THE nature of instinct or special sense which naturalists have studied in animals is, without doubt, one of the most puzzling and complex problems confronting the modern physiologist.

It reflects, under its most strange and least explored aspect, the whole problem of life. Yet, in spite of great difficulties in the field of observation, accurate data on this subject have been recorded from time to time. In this matter, the experimental method is practically restricted to direct observation, and more often than not laboratory experiments are out of the question.

Various hypotheses have been advanced to explain the observed and controlled results, but it would seem that up to the present no general theory has yet been enunciated which would cover all the available data and at the same time give a logical and comprehensive explanation.

In this connexion the uninterrupted progress of science is suggestive of certain new ideas which have enabled me to elaborate my theory of the origin of life and of radiation in relation to living beings, forming the subject of the present work which began to appear from 1923 onwards in various periodicals.

The Instinct of Orientation

At the outset I devoted my attention to investigating the causes of the ease with which certain animals succeeded in finding their bearings so unerringly during the longest voyages. Such are carrier pigeons, which return to their dove-cot after having flown a few hundred miles. Another example is migrating birds, which fly in a straight line day and night, speeding across the seas towards a definite destination that they cannot possibly perceive, partly because of their limited visual powers and partly because of the curvature of the earth's surface. They emigrate to feed on insects that they can no longer find in our latitudes at the approach of winter.

Some say that this is sheer instinct, while others prefer to call it special sense, but neither term explains the riddle. I hold that in science nothing should be mysterious. Such terms as instinct and special sense merely mask our ignorance and it should be possible to account for everything.

It seems more and more evident, as the following observations make it clear, that the sense of direction originates from special radiations of ultra-short wavelength, emitted by the birds and insects themselves.

Carrier Pigeons. We have all heard of the truly wonderful powers of orientation possessed by carrier pigeons. Although this faculty is innate it nevertheless requires a certain training before it is fully developed.

After the bird has risen in the air and circled round a few times, this faculty of orientation enables it without hesitation, even at night, to fly towards its dove-cot, which is sometimes far away.

I have noticed the prevalence of this phenomenon and have ventured to give an explanation of it in the present work : all birds about to undertake long migration voyages (wild ducks, wild water-fowls, swallows, etc.) invariably describe, like carrier pigeons, a series of orbits in the air before starting on their final flight.

A most interesting observation made on July 2nd, 1924, at the radio station of Paterna, near Valencia (Spain), came to my notice. A flock of pigeons had just been released near an aerial of this station at the time of transmission. It was then observed that these birds could not manage to find their bearings and kept on flying in a circular fashion, as if completely disorientated. This experiment was repeated several times and always produced the same result, that is to say the disappearance, or rather a very marked perturbation of the sense of direction in carrier pigeons under the influence of electromagnetic waves.

These experiments were taken up again at Paterna, at the radio station of Valencia, under the control of the Spanish military authorities,[1] and also at Kreuznach (Germany). These fresh experiments fully confirmed my views concerning the influence of Hertzian waves on the instinct of orientation.

A Spanish scientist, M. J. Casamajor, wrote a detailed report on the Paterna experiments. The Spanish carrier pigeon service installed a military carrier pigeon station at Valencia, at a distance of about 8 kilometres from the radio station of Paterna. At the time of the experiment in question pigeons were released one by one at regular intervals of three minutes near the station while transmission was taking place continuously. It was observed that all the pigeons began to fly by circling round for some time, but without succeeding in finding their bearings as they usually do after having flown round a few times. In spite of a change of wavelength in the course of transmission, no return to the normal condition was observed, and so long as transmission occurred, and it lasted more than half an hour, no pigeons succeeded in flying in a definite direction. It is important to note that barely a few minutes after the transmission was over the released pigeons flew towards their dove-cot without the least hesitation, even those which had taken part in the first experiment.

Another series of experiments which took place on November 7th, 1926, in the same locality produced the same result.

The original experiments at Paterna put investigators on their mettle, for they could not understand the relation existing between the instinct of pigeons and the transmission

[1] This was written in 1925. (Translator.)

of electromagnetic waves. The German technicians hastened to verify and control Casamajor's observations. In March, 1926, they initiated a series of experiments similar to those carried out at Kreuznach. The conditions, however, were different and more rigorous. A site was chosen so that the dove-cot and the radio station were diametrically opposed. Consequently this station was situated exactly as the crow flies on the course that the pigeons were bound to take. On arriving near the radio station it was noticed that the pigeons changed their flight, were losing their bearings and appeared to be definitely disorientated. They did not succeed in resuming their course towards the dove-cot until their flying had brought them outside the intense electromagnetic field surrounding the aerial of the radio station.

It is noteworthy that the simplest explanation of this phenomenon does not seem to have occurred to any of the Spanish, French and German experimenters, namely that of electromagnetic induction on the pigeons' directive organs. They were all baffled by the significance of the phenomenon which they attributed to a curious anomaly that they could not explain.

Nocturnal Birds

The Bat. The observations made on carrier pigeons appear to hold good for nocturnal birds also. It seems obvious, *a priori*, that the sensibility of these birds to electromagnetic waves in general is different from that of diurnal birds by virtue of their special adaptation to light or darkness. These two species of birds, however, show a common feature, they feed on the same insects.

We are led to believe, as we shall see later, that they are attracted to their prey by radiations emitted by these insects. There is little doubt that daylight has an influence on the propagation of these radiations. If sunlight absorbs them, as it does in the case of wireless waves, nocturnal birds (various species of owls) should go hunting at night because their sensibility to reception, so far as these radiations are concerned, is less developed than that of diurnal birds. Conversely, if sunlight increases the amplitude of radiations, as seems to be the case for waves measuring several metres, then it is the excess of intensity of the radiations which would prevent nocturnal birds to go hunting during the day.

In this matter of sensibility of reception to special radiations, one is justified in assuming the existence of correlative differences in the organs of sight, as observed in diurnal and nocturnal birds. Among nocturnal birds, let us take the bat as an example. It is commonly believed that it is to the acuity of the senses of hearing and smell that the bat owes its ability of approaching its prey whose least movements it can detect, thanks to the vibrations of the air reaching its ears. This hypothesis may be admissible under certain conditions such as the calm atmosphere of the countryside. In Paris I have often watched bats from my balcony, on racing days, amid the uproar of a great crowd and the noise of thousands of cars setting up vibrations in the air, saturated with the products of petrol combustion. Amid this deafening din and vitiated atmosphere it is neither the sense of smell nor that of hearing that guides the bat straight towards insects (cockchafers, moths, etc.) which they catch as easily as in the undisturbed silence of the countryside.

The bat is thus most probably attracted to these insects by the radiations they emit, which are not influenced by noise nor by petrol fumes.

Lemmings. This is another extraordinary example, the lemming, a kind of field-mouse whose habitat is in Scandinavian regions. The famous Swedish naturalist, Linnæus, gave an account of their peculiar expeditions.

"At the approach of severe cold weather and sometimes without any apparent reason, lemmings leave their natural habitat in the high mountains of Norway in order to make a long voyage towards the sea. The emigrating throng, consisting of myriads of individuals, trots in a straight line across all obstacles without ever letting itself be diverted from its goal. While proceeding in Indian file they trace rectilinear parallel furrows, two fingers deep and several yards apart. They devour anything obstructing their passage, such as herbs and roots. Nothing diverts them from their route. If a man should come across their path they run between his legs. If they meet a haystack, they gnaw their way through; if it should be a rock, they go round it in a semicircle and resume their straight course. Should a lake impede their progress they swim across it in a straight line, whatever its size may be. Is a boat in the way? They climb over it and dive into the water on the other side. A strong current in a river does not stop them, even at the risk of annihilation." [1]

Is it possible that these animals are guided in their straight course by their sense of smell or hearing? They perceive smells and noises coming from all directions. Is it not simpler to suggest that these lemmings, although feeding on roots and seeds, and needing an occasional addition of small fishes, travel towards the sea, guided by the radiations emanating from the shoals of fishes upon which they feed? Furthermore, glow-worms, micro-organisms in decomposing meat, fire-flies, etc., emit luminous radiations. And so, too, with certain animalculæ whose presence in innumerable masses makes the sea phosphorescent. It is also common knowledge that certain fishes known as torpedo-fishes, give off electricity.

Thus an elementary intuitive generalisation would seem to establish the fact that certain animals emit radiations which we cannot perceive, but whose effects are far-reaching.

Rôle of Semi-circular Canals in Birds and of Antennæ in Insects

Some naturalists have stated that the semi-circular canals of the ear, in many species, are endowed with special directing properties. If these organs are removed, the operated birds invariably lose their sense of equilibrium and turn round and round, as though stupefied and incapable of taking a definite direction. Assuredly here is an interesting observation. But another observation of the highest importance has been made by scientists. The fluid contained in the semi-circular canals would appear to be particularly sensitive to the influence of an electromagnetic field while the walls

[1] In his text-book on Zoology, Sedgwick wrote "The Scandinavian lemming migrates in a straight line in enormous herds, crossing all obstacles till it reaches the sea into which it plunges in the continuance of its wandering and is drowned." (Translator.)

of the canals consist of insulating material. Now, any wireless transmitter creates a variable electromagnetic field whose action makes itself felt at considerable distances. In view of this fact we may well ask ourselves whether a great number of living creatures do not obtain their bearings through the agency of waves similar to those transmitted by radio stations.

The semi-circular canals are susceptible of playing the rôle of a radiogoniometric receiver.[1] The very conformation of the semi-circular canals appears to support this hypothesis. They are arranged in three planes, each of which is at right angles to the other two so that in the semi-circular canals the three planes of space are represented. Such a scheme constitutes a system of co-ordinates,[2] necessary and adequate to determine the position of a point in space, or, in the case under consideration, the position of a bird in the atmosphere, or yet the position of an insect in relation to the bird (Fig. 1).

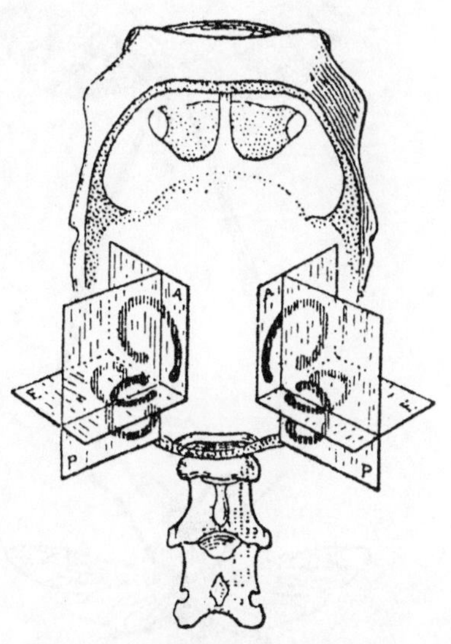

FIG. 1. Schematic diagram of semi-circular canals.

A, plane of anterior canal; P, plane of posterior canal; E, plane of horizontal canal (after Ewald).

Animals in general, and birds in particular, do not move in a horizontal plane but in a three-dimensional space and the semi-circular canals have been devised accordingly.

The conducting fluid contained in these canals constitutes a directional receiving circuit completed by an accessory circuit in the form of a pliable spiral (self-inductance and tuning capacity).

In the strange world of insects many of them possess

[1] In wireless, a radiogoniometer is a kind of directional receiving apparatus. (Translator.)

[2] A system of lines by means of which the position of a point is determined. (Translator.)

minute antennæ enabling them to follow their course in a straight line towards relatively distant points. Nature does nothing in vain ; these antennæ would seem to exist only for the purpose of receiving radiations (Fig. 2).

The similarity between the antennæ of insects and the aerials of radio stations is striking, but this similarity, how-

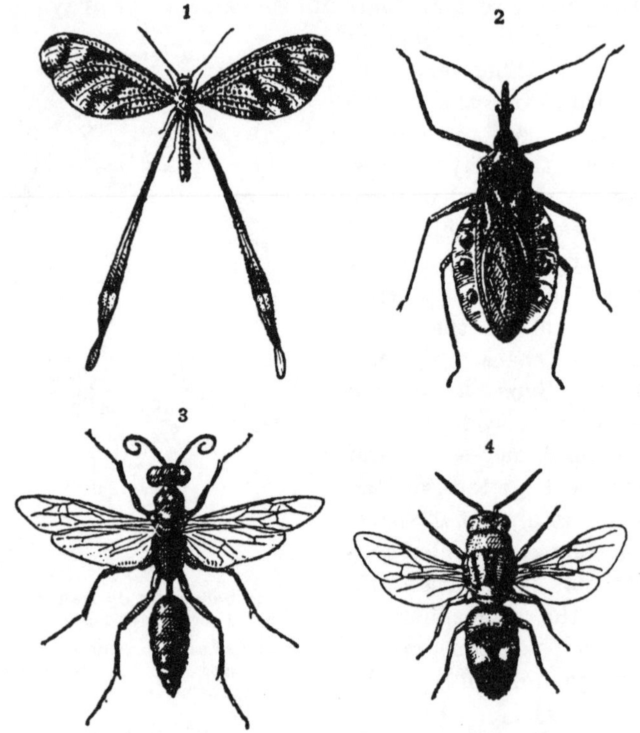

FIG. 2. Four species of insects with characteristic antennæ.
1. *Nemoptera lusitanica*. 2. *Eulyes melanoptera*. 3. *Chlorion lobatum*. 4. *Euchrœus purpuratus*.

ever, is not as simple as it would appear at first sight. Owing to their relatively considerable dimensions in relation to the emitted radiations, the antennæ of insects function in the manner of complex oscillators vibrating with the frequency of harmonics of a far higher scale than their fundamental wavelength.

Nocturnal Experiments with the Great Peacock-butterfly. Let us consider, by way of example, the bombyx, in the light of observations made by Fabre in his work entitled "*Mœurs des insectes.*" In the laboratory, soon after the emergence of a female from the chrysalis, Fabre observed that, at night, a whole swarm of males invaded the place, which leads us to suppose that this female was endowed with a certain "nocturnal capacity." Fabre also pointed out the difficulties of access to his laboratory surrounded by a multitude of trees. In spite of these obstacles the males always succeeded in reaching the female. The following day the same phenomenon was observed ; it all seemed as if the sense of smell had been guiding the moths. Fabre then gives an account of experiments which shatter this hypothesis.

In the first place, the moths of this species, known as the Great Peacock, are well-nigh impossible to be found under normal circumstances. Thus the males must have come from a very distant site. Sound, light and the sense of smell are out of the question, for the moth makes straight for the cage in spite of a variety of scents intentionally diffused by the experimenter in order to lead the insects astray. The factor of place-memory may be ruled out as irrelevant.

Fabre also remarked that the moths were travelling in the same direction as the wind. It follows that if they had been guided by the sense of smell they would have had to soar with the wind in order to catch the scented air.

Diurnal Experiments with the Oak-bombyx. In order to ascertain the influence of sunlight Fabre experimented in full daylight by studying the habits of the oak-bombyx, whose diurnal activities are more pronounced. But this insect, like the Great Peacock, is not to be found in the region where Fabre was working. How are we to account for the fact that it was able to come from its distant habitat ? The males hurried along and found the female locked up in a drawer or under a framework covered by a cloth, in spite of nauseating effluvia emanating from all sorts of odoriferous substances placed there by the experimenter.

According to Fabre, the following experiment would seem to confirm the olfactory sense hypothesis.

"I placed the female in a bell-glass and gave her a slender

oak twig with withered leaves as a support. The glass was set upon a table facing the open window. On entering the room the moths could not fail but see the prisoner, as she was placed directly in their way. The tray, containing a layer of sand, where the female had passed the preceding morning and night, covered with a piece of wire gauze, was in my way. Without premeditation I placed it at the other end of the room, on the floor, in a corner where but little light could penetrate, about ten steps away from the window.

The outcome of these preparations completely upset my notions. None of the arriving insects stopped at the bell-glass where the female was plainly to be seen in full daylight. They passed on as though indifferent. Not a glance, nothing to put one on the track. They all flew to the further end of the room into the dark corner where I had placed the tray and the bell-glass. They alighted on the wire dome. . . . All the afternoon, until sunset, the moths danced about the empty cage a saraband which the real presence of the female would normally evoke. . . . Finally they departed, but not all. There were some who would not go, as if held there by some magical force. Truly a strange result. The moths collected where there was apparently nothing. . . . What had deceived them ? All the preceding night and all the morning the female had remained under the wire-gauze cover, sometimes clinging to the wirework, sometimes resting on the sand in the tray. Whatever she touched, above all, apparently, with her distended abdomen, was impregnated, following a long contact, with certain emanations. This was her lure, her love-philtre. This it was that revolutionised the insect world. The sand retained these emanations for some time and diffused the effluvia in turn. Thus it is the olfactory sense that guides the moths and warns them far off. . . . The irresistible philtre requires time for its elaboration. I imagine it as an exhalation which is gradually given off and saturates whatever is in contact with the motionless body of the female. . . . With these data in hand and unexpected information resulting from them, I varied the experiments, but all pointed in the same direction. In the morning I placed the female under the wire-gauze cover ;

for support an oak twig was provided. There, motionless, as if dead, she lay for hours, buried under a cluster of leaves which would thus become impregnated with her emanations. When the hour of the daily inspection drew near, I removed the twig and put it on a chair not far from the open window. I left the female under the bell-glass, plainly exposed on the table in the middle of the room. The moths arrived as usual . . . they hesitated . . . they were still searching. Finally they found something, and what did they find ? Just the twig. . . . With their wings rapidly fluttering they alighted on the foliage exploring it all over, probing, raising and displacing it until at last the twig fell on the ground. Nevertheless, they continued probing between the leaves."

From his experiments Fabre concluded that these moths were endowed with a sense of smell very different from ours and characteristic of their species.

Fabre's conclusion fails to satisfy me.

The act of smelling is dependent on material particles which excite the olfactory sense, but the diffusion of these particles is limited to a short radius in the atmosphere. Thus it is not due to these particles that the moths were enabled to fly long distances.

I thought it fit, therefore, to repeat these experiments.

In my view, what attracts the males towards the female in the case of the Great Peacock and the Bombyx, is not the splendour of her coloured mantle and her velvet wings, nor is it the odoriferous particles. It is rather the infinitesimal particles given off by her ovaries, micro-organic cells radiating according to a scale of determined wavelengths and exciting in the males the desire of procreation.

This hypothesis is confirmed by the following experiment which I carried out myself.

New Experiments with the Oak-bombyx. After the emergence of the female from the chrysalis, a host of males rushed from all directions. After having left during the night this female lying on a leaf of cotton wool, I removed her the following day at noon. Then I placed, at a distance of about 5 metres from the female, the cotton-wool leaf on which the males came to rest again.

I repeated this experiment after having this time dipped

the cotton wool in a solution of pure alcohol, and I observed that the males stopped coming. The same result was obtained when corrosive sublimate was used instead of alcohol. Now, neither pure alcohol nor corrosive sublimate could have had the least effect on the odoriferous effluvia. On the other hand, these solutions had destroyed by sterilisation the living cells which gave off the radiations that attracted the moths.

Burying-beetles (*Necrophorus*). The activities of these beetles on the decomposing bodies of dead rats and birds also appear to confirm my theory.

As some naturalists have remarked, these insects play a hygienic part in the economy of nature, in fields and woods ; they scavenge upon death for the benefit of life. They belong to a certain species of insects which attack dead bodies and devour them until they have restored into the cycle of life this inanimate organic matter. The burying-beetle is essentially a grave-digger, sometimes travelling long distances to reach the dead bodies of rats and birds which it buries by degrees into the earth so that they may ultimately serve as food for its offspring destined to be born on the same site.

The extraordinary social life of these beetles might be described at length. Let us confine ourselves to a characteristic which is relevant to our theory, the fact that they know how to direct themselves across great distances towards the dead bodies of rats and birds.

Is it likely that they are guided by the sense of smell ? If dead bodies give off odours, the odoriferous particles cannot be diffused beyond a range of a few metres. This hypothesis is inadmissible, in the case of burying-beetles, as in other cases, in view of the great distances that have to be covered.

It is also important to observe that the beetles do not appear until eight to ten days after the death of birds or rats, when their bodies are in a state of decomposition.

It would seem, therefore, that it is the micro-organisms arising out of this decomposition and oscillating according to a determined scale of wavelengths, which direct the burying-beetles or their offspring towards their food.

CHAPTER II

AUTO-ELECTRIFICATION IN LIVING BEINGS

Electrification by Friction of Wings in the Atmosphere—Influence of Electrical Capacity in Birds—The Rôle of Orientation in the Flight of Birds—Explanation of Migration—Extension of Principle to Wingless Animals.

Electrification by Friction of Wings in the Atmosphere. Simple experiments have confirmed the following hypothesis that I had previously formulated ; living beings moving in the atmosphere, notably insects and birds, are capable of taking electrical charges, often at a very high potential.

In imitating the flight of a bird in order to study the effects produced by the friction of its wings against the air, as, for example, by shaking a duck's wing before a radium electrometer after having taken care to insulate myself from the earth by means of two ebonite discs of 2 cm. thickness, I have been able to measure a charge of static electricity of an approximate tension of 600 volts. This tension increases as the earth level becomes further distant from the experimenter.

These experiments put an end to all the controversies that have raged for the past fifty years among investigators (naturalists, entomologists, ornithologists, hunters, etc.) on the subject of the migration of birds in general, and of their direction in relation to that of the wind in particular. It is only fair to state that the majority of observers have admitted that their conclusions were, after all, but approximations, the solution of the problem thus remaining to be found.

As I have already stated, all living beings emit radiations. But, as far as the reception of these waves is concerned, birds which feed while flying have a far greater capacity and sensibility than animals that are restricted to moving on the earth's surface.

We know that the electric potential of the terrestrial atmosphere increases with height at the rate of 1 volt per cm. Thus at a height of 1,000 metres there is a potential difference of 100,000 volts in relation to the earth's surface. This increase of potential with height accounts for the formidable charges observed in certain aerial metallic tracks situated in mountainous regions. It also accounts for those luminous brushlights which, in the calmest atmosphere, alpinists have observed being shot off their ice-axes at a high altitude such as the summit of the Wetterhorn in the Bernese Oberland (3,703 metres).

Moreover, it has been observed that all birds about to undertake a long migration voyage (wild ducks, pigeons, swallows, etc.) start by rising in the air, then describe a series of numerous orbits before taking their final departure.

Why do they fly in this manner ?

Judging by what we have just learned about the instinct of orientation, we may assume that in describing such orbits the birds avail themselves of a useful process to ascertain the various directions of atmospheric waves by means of their natural radiogoniometer (radio-direction finder), consisting of the semi-circular canals.

It is highly probable that the purpose of these preliminary manœuvres lies essentially in the necessity, imposed on the birds, to obtain the indispensable electric tension in order to detect insects or other prey they are searching for, which are actually thousands of miles away.

As a case in point let us suppose that if, to the atmospheric potential generated by altitude, say 50,000 volts for an ordinary flight at a height of 500 metres, we add the potential developed by friction of the bird's wings against the wind, say 25,000 volts, we arrive at a total of 75,000 volts.

Influence of Electrical Capacity in Birds. It is worthy of note that electric tension during a bird's flight varies in direct ratio to the resistance of the wind. The stronger the wind, the greater the electric tension acquired by the bird. The weaker the wind, the more this tension diminishes.

Again, when the bird flies in a straight line, it encounters on its path winds of variable intensity coming from all directions. This electric tension may thus be regulated by the bird which simply flies high or low according to the strength and direction of the wind. If, in the course of a flight against the wind, the electric tension should rise from 75,000 volts to 100,000 volts, the bird must come down a distance of 250 metres in order to bring the tension back to the former figure. At this new altitude the bird will find in the atmosphere an electric tension which, added to that generated by the friction of its wings against the wind, will give it the tension of 75,000 volts which is both sufficient and necessary for continuing its flight. On the other hand, a higher tension would prove detrimental.[1]

Thanks to this means of regulating its electric tension by varying the flying level from the earth's surface, the bird, together with the underlying soil, constitute an actual air condenser.

The bird thus possesses a kind of complete wireless apparatus since the semi-circular canals, in communication with his brain, and under the influence of electricity, play the part of receiver.

Just as for picking up wireless waves emitted in America the operator regulates the mechanism of his receiving apparatus by modifying with a variable condenser the capacity of his aerial in relation to the earth, so the migrating bird regulates his own electrical capacity by flying either high or low.

The Rôle of Orientation in the Flight of Birds. A Belgian entomologist, Dr. Quinet, after having made observations for thirty years, states that he has invariably " seen " birds flying against the wind. The theory that has been put forward in this work provides a simple explanation of this phenomenon. When they fly against the wind, birds are compelled, in order to lower their electric tension, to come down to low altitudes which enables the observer to see them clearly. But when birds fly with the wind they rise to a considerable altitude so as to obtain the charge of atmo-

[1] It is known that the electric tension of the atmosphere is proportional to the altitude; on the other hand, the electrical capacity of the bird in relation to the soil is, in the first approximation, inversely proportional to the altitude, The result is that the product of these two quantities, which is the electric charge of the bird ($Q = CV$) is constant. This electric charge appears to be a constant for any given bird.

spheric electricity which is indispensable to them. In this case the birds remain invisible to the naked eye.

This theory also furnishes an explanation of the observations made by Ternier and Masse, Cathelin and Aubert, when they stated having " heard " and " seen " migratory birds flying at great heights with the wind or against a light breeze.

All these different observations, far from excluding one another, combine to confirm my theory.

Explanation of Migration. On the subject of migration of birds and the means employed by them to that end, naturalists have advanced a great variety of hypotheses. Some have attributed the migratory instinct to an exceptionally acute sense of sight, while others have imagined the existence of an extremely sensitive hearing thanks to a kind of microphonic apparatus. There are yet others who have supposed that the birds were endowed with a highly developed olfactory sense enabling them to detect effluvia which escape us. There are also those who have invoked an electromagnetic action, localised in the atmosphere ; and lastly there is the hypothesis of place-memory.

The majority of observers appear to have preferred the instinct or special sense hypothesis.

All these theories do not explain why, for example, the falcon rises facing the wind before pouncing on its prey, which it does not seem to perceive standing close by ; nor why sterns and seagulls perform a series of circular manœuvres in the air, while facing the wind, before alighting to fish in the waves. Nor yet do these theories explain a host of analogous facts.

The theory of auto-electrification alone, stating that the bird is able to detect radiations emitted by the living things upon which it feeds, may be said to explain these phenomena that have hitherto remained so mysterious.

Extension of the Principle to Wingless Animals. Although animals that live in close contact with the earth's surface electrify themselves less easily than birds and insects, it is nevertheless a fact that they are endowed with a certain degree of receptivity which enables them to detect radiations but only within a very restricted radius. Thus the horse is capable of finding his way to the stable within a radius of 10 kilometres. The dog "detects" his master within reasonable distance. Lemmings travel towards the sea from the far distant mountains of Norway. And the same principle applies to all animals possessing a tail, for they all electrify themselves by waving their tail in the air. It should also be noted that the tail of animals producing auto-electrification serves both as an antenna and an aerial. Moreover, the tail is in direct connexion with the most important nervous centres.

CHAPTER III

UNIVERSAL NATURE OF RADIATION IN LIVING BEINGS

Fundamental Principles—Nature of Radiation in Living Beings—The Glow-worm.

Fundamental Principles

As a result of numerous observations and experiments I have formulated the following four principles :

1. Every living being emits radiations.[1]

2. The great majority of living beings—with very few exceptions—are capable of receiving and of detecting waves.[2]

3. Any flying creature, that is to say, capable of leaving the earth's surface (bird, winged insect) possesses a high capacity of wave-emission and reception, while animals that are unable to fly have a far lesser capacity in the same direction.[3]

4. The influence of sunlight on the propagation of waves is the determining factor in causing certain birds and insects, whose receptivity is specific, to fly and to feed at night, whereas others whose receptivity is normal, function, so to speak, in the daytime.[4]

[1] This first principle is the keystone of the theory. Evidence of its validity is given in the following chapters.

[2] The second proposition is a natural corollary of the first. The work of physicists on wave propagation has shown that any transmitting system is susceptible of receiving waves and of transmitting them. Indeed, every radiating system can both emit and transmit.

[3] The third proposition is of a somewhat intuitive order and is based upon what everyone knows on the propagation of radiations. Absorption of waves is greater in the soil than in the atmosphere. High aerials are better than low ones for emitting and picking up waves. It follows, therefore, that flying creatures are better equipped than non-flying ones for emitting and receiving radiations.

[4] The fourth proposition accounts for the differences observed as much in the organs as in the habits of diurnal and nocturnal animals respectively. All observations on Hertzian waves show the undesirable influence of solar radiation on the propagation of waves. But we are not yet in a position to know definitely to what extent and in what way this influence is exerted on ultra-short waves. As far as waves of several hundred metres are concerned and also longer waves, sunlight has a very marked weakening effect. As for waves under 100 metres, the reverse effect occurs, complicated by the phenomenon of scintillation.

We may now adapt these conclusions to living beings whose radiations are equally influenced by sunlight.

As the modern tendency is to reduce all physical phenomena to unity by bringing into play the full range of waves, it is perfectly logical to assume that certain animals act as transmitters and receptors of radiations. It seems almost certain that the majority of insects and birds give off radiations, and are also sensitive to the influence of waves which enable them to find their bearings. In any case, these creatures find their way under the influence of waves, and this orientation is automatic.

When, in 1923, I conceived my theory, these principles could only be considered as a possible hypothesis. But as a result of all the observations and experiments I have made since then this hypothesis seems to me to have gained a greater measure of clearness and validity.

Nature of Radiation in Living Beings

In order to understand fully the rôle and nature of radiations emitted by living beings, it may be instructive to look back and recall the history of the discovery of electromagnetic waves. The existence of these waves was not generally known until an apparatus had been devised to render them perceptible to our senses. The greatest claim to fame on the part of Hertz, Branly, Marconi, and many other technicians and amateurs, lies essentially in having invented an apparatus which, independently of all theories on the nature of radiation, makes these waves easily perceptible, even across great distances.

The recent discoveries of certain kinds of radiations—wireless waves, X-rays, radio-activity, cosmic rays—have but slightly lifted the veil of mystery concealing from our senses whole gamuts of waves which elude direct perception.

Is it not possible that we are surrounded by other radiations, imperceptible to us, because we do not possess the necessary apparatus capable of revealing them to our senses?

If we admit that birds emit and detect radiations imperceptible to us, the terms instinct and special sense employed to explain certain characteristics become clear immediately, and assume a precise significance. The sense of orientation in birds, and in animals generally, explains itself at once.

Just as a ship lost in the fog tries to ascertain by means of a radiogoniometric apparatus the direction of the Hertzian beacon sending off electromagnetic waves, so, too, the animals and insects in question try to pick up radiations emitted by living beings and plants which have a definite interest for them. Their orientation is subsequently determined by the bearings obtained.

But it may be objected that space would then be riddled with innumerable radiations. How would it be possible for these creatures to detect them ?

The answer is simple. Discrimination is easily effected thanks to the diversity of frequencies which characterises these radiations. We shall see how this is accomplished presently.

What is the organ which enables an animal to pick up these waves and to detect them while also rendering them perceptible to their senses ? My firm conviction is that this organ is the semi-circular canals of the ear whose fluid is sensitive to electromagnetic fields, thus enabling animals to be aware of the vibrations they are searching for.

We may now examine more closely the functions of the semi-circular canals by studying the modalities of their configuration in different living species.

The invertebrates do not possess any semi-circular canals, but only membranous vesicles which take their place and have similar functions. Yves Delage mentions the case of the octopus which is still able to swim after being blinded, but turns round its longitudinal axis or plane of symmetry when the vesicles which control its faculty of orientation have been destroyed.

After the destruction of both labyrinths aquatic animals and notably frogs, can no longer swim nor jump in a straight line. It should also be noted that lampreys, which have only two pairs of canals, can only move in space in two directions ; that Japanese mice (dancing mice) which only possess the superior vertical canals, can move only in one direction, right or left, and are moreover incapable of moving straight ahead or in a vertical direction. These rodents, as E. de Cyon has shown, know only one space of one dimension.

The majority of the vertebrates possess semi-circular canals arranged in three planes in space. This assemblage of three canals, each of which is at right angles to the other two, constitutes the labyrinth which is completed by more or less developed organs: the vestibule and the cochlea.[1]

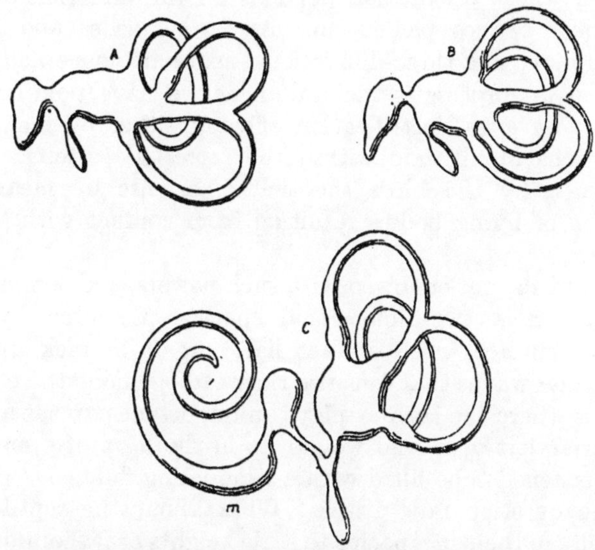

FIG. 3. Diagram of semi-circular canals in different species of vertebrates· A, fishes; B, birds and reptiles; C, mammals (after Waldeyer).

It should be noted that apart from the three semi-circular canals being disposed at right angles to one another, these organs are differentiated by features corresponding to the particular needs of each species. Fishes obtain the necessary electric tension by friction resulting from the impact of their bodies against water and by swimming nearer to or further from the earth's level. Similarly birds acquire auto-electrification by friction of wings against the air and by varying the altitude in the course of their flight. Mammals which cannot avail themselves of either of these auto-electrification processes are in need, in order to pick up waves, of a special directing apparatus represented by the accessory spiral, *m*.

Now, whereas the cochlea is highly developed in mammals it is practically absent in fishes, reptiles and birds (Fig. 3).

How may we account for this difference? Is the presence

[1] In physiology the *labyrinth* is a name given to the series of cavities of the internal ear. It comprises the vestibule, cochlea and semi-circular canals. The *vestibule* is an oval cavity of the internal ear which forms the entrance of the cochlea. The *cochlea* is a cavity of the internal ear resembling a snail-shell. Helmholtz was of the opinion that it served the purpose of analysing sound waves. (Translator.)

of the cochlea in mammals related to a special sense which is absent in birds and fishes ? I believe that, from the point of view of my theory, the question is susceptible of a very simple and general explanation. We have already seen that the semi-circular canals function as a radiogoniometric system whose orientation depends on the direction of the particular waves picked up. As far as fishes and birds which move in a three-dimensional space are concerned, this picking up process is facilitated, as we have pointed out before, by auto-electrification effected either by means of variations of altitude within the terrestrial electric field, regulated by the birds themselves, or else by means of friction of living bodies resulting from contact with air or water.

Mammals, not endowed with such powers, and confined to moving in a two-dimensional space represented by the earth's surface, need an auxiliary organ to pick up the particular waves that sensitise their radiogoniometric canals. This is where the cochlea plays an important part, as a kind of aerial, left open and wound up in the form of a more or less flattened tube filled with a conducting fluid.

The question now arises "What about the reptiles ?" In spite of their incapacity to scale heights or fathom depths why are they not brought in the same category as mammals and why are they devoid of a cochlea ?

The answer will be apparent to anyone who has observed the movements of reptiles. If, by chance, on a warm summer's day, you have the opportunity of seeing an adder, for example, you may observe that while resting, its long articulated body is arranged somewhat in the form of a flattened coil. This apparent state of repose or sleep which the snake seems to have assumed, is in reality a state of subconscious watching. The adder watches ; the harmonious winding of its body is a small receiving apparatus which to a great extent makes up for the absence of a diminutive cochlea in the labyrinth containing the semi-circular canals. If an owl, or any other diurnal bird of prey should venture to approach the snake or if a harmless green frog, an easy prey, should go near it, this improvised receiving apparatus, formed by the adder's body, will immediately warn the snake, which will be prepared either for attack or for escape. This would seem to prove the needlessness of a specific spiral apparatus for picking up waves.

Thus, once again, we have a confirmation of the ancient dictum : " Nature does nothing in vain " ; and there is no reason why a useless organ should be preserved when Nature finds a better substitute for it.

What then are these radiations emitted by living beings ? Like all other known radiations, they are characterised by their wavelength. Our present task is to consider the range of wavelengths that comprises these radiations.

The Glow-worm. At the outset, let us show by a concrete example that it would be absurd to deny the principle that living beings emit radiations. This negation is obviously futile, as all available data formally contradict it.

No great mental effort is required to think of an insect which emits luminous radiations, I mean the glow-worm.

What is the glow-worm ? An insect that remains more or less constantly in a luminous state. Experiments have shown, by direct observation, that the eggs of the glow-worm are spontaneously luminous and that this characteristic light is transmitted without a break from generation to generation.

What then is this radiation of the glow-worm ? Nothing but radiations of ordinary light, but filtered and giving a special luminous spectrum that may be observed with the spectroscope. Hence if we perceive the luminescence of the glow-worm, it is primarily because it is due to a luminous radiation, emanating from cells, certain molecules of which vibrate with the same frequency as light which we can perceive immediately because it affects our visual sense.

Why then should we admit the possibility of the glow-worm emitting luminous radiations while refusing to admit the possibility of other insects emitting different types of radiations beyond the range of luminous ones, and consequently imperceptible to our senses ?

Such an attitude is reminiscent of St. Thomas, for we insist on seeing the radiations before believing in their existence. But we know that in the incommensurable range of vibrations, only the luminous octave is visible to us.

There is no gainsaying this, and the mystery of the cases under consideration vanishes if we admit that the fact of emitting radiations is a universal property of living matter, just as it is becoming more and more evident that radio-activity is a universal property of inanimate matter. We may ask ourselves whence comes the energy necessary for radiation. We shall see later how this question may be answered in its generalised form and also in regard to all living beings. In any case, it seems inconsistent not to concede to other living beings what is conceded in the particular case of the glow-worm.

The full range of radiating properties of living beings does not manifest itself to our senses any more than the complete gamut of electromagnetic waves.

Let us humbly remind ourselves that the human body has but very small windows looking out upon the incommensurable range of an ocean of radiations. Our senses can reveal to us but a few octaves. The scanty knowledge we have concerning radiations of living beings must suffice to guide us in the study of the whole range.

We have drawn attention to the luminescence of the glow-worm which emits a cold light, or nearly so. It is hardly necessary to add that all animals with a constant normal temperature or a temperature higher than that of the ambient atmosphere, emit calorific radiations, *i.e.*, warm radiations.

Before formulating a general theory and dealing with the problem of energy, let us say a few words on radiations in general, and especially on electromagnetic radiations with which modern science has made us familiar. These radiations constitute the basis of the most important phenomena in physics. The propagation of sound waves through matter is effected against a certain amount of resistance while electromagnetic waves traverse the most tenuous space filled only by the all-pervading ether. Among such waves we find wireless waves, calorific waves, luminous waves, actinic waves, X-rays and penetrating waves (cosmic rays).

CHAPTER IV

ON RADIATIONS IN GENERAL AND ON ELECTROMAGNETIC WAVES IN PARTICULAR

Nature and Characteristics of known Radiations—Table of Radiations—Electromagnetic Waves—Rôle of Self-induction and Capacity—The Oscillating Circuit—Natural Period and Resonance—Explanatory Analogies concerning Electrical Oscillations—Ultra-short Waves.

Nature and Characteristics of Known Radiations

It is generally known that a radiation is a disturbance of the ether travelling at the velocity of light, that is to say 186,326 miles per second. The range of known radiations comprises wireless waves, calorific, luminous, chemical radiations, X-rays, gamma-rays of radium and cosmic rays. These various radiations differ from one another only by their frequency, that is to say by the number of oscillations per second which characterises them. The wavelength is the distance covered by the wave per second in the course of its propagation. The higher the frequency of radiation the shorter is its wavelength. The process of radiation does not involve transport of matter or emission of particles; it is essentially the propagation of a disturbance occurring in the ether.

Such are the main principles of the theory of radiations governing modern physics.

The table on p. 56 represents the complete scale of electromagnetic waves with their respective wavelength and frequency.

According to Clerk Maxwell who conceived a famous theory of light, luminous radiation is of a purely electromagnetic nature. As electromagnetic waves are now familiar to all, we propose considering them at some length. This apparent digression is necessary in order to get a clear grasp of the technical details that will be given later in

TABLE OF ELECTROMAGNETIC WAVES

Type of Wave.	Wave-length.	Frequency (Vibrations per second).
WIRELESS WAVES .	30,000 metres to a few millimetres.	10,000–50 milliards
INFRA-RED WAVES . (Calorific rays).	314μ–0·8μ	1–375 trillions
LUMINOUS WAVES . (1 octave).	0·8μ–0·4 μ	375–750 trillions
ULTRA-VIOLET WAVES.	0·4μ–0·015μ	750 trillions–20 quatrillions
X-RAYS . . . (12 octaves).	0·015μ–0·0000057μ	20 quatrillions–60 quintillions
RADIO-ACTIVITY. .	0·0001μ–0·000002μ	3–150 quintillions
COSMIC WAVES (Penetrating radiation).	0·0002 A.U.	—

Greek letter μ (micron) = a thousandth part of a millimetre.
A.U. (Angström unit) = a ten-millionth part of a millimetre.

This table covers about 60 octaves out of which the human eye can detect only 1 octave. All these rays are believed to have certain common characteristics. They are generated by moving electric charges and propagated without any material medium. They are also supposed to travel with the same velocity of about 186,000 miles per second.

connexion with my theory of radiation of cells and of living beings. Moreover, anyone should be able to follow easily the explanations and analogies given in this chapter concerning oscillating circuits and high-frequency currents. Readers capable of understanding technical accounts of electromagnetic waves may find useful information in the footnotes on self-induction and capacity in the oscillating circuit.[1]

[1] *Electromagnetic Waves.* The phenomena associated with electrical oscillations cannot be fully understood until a certain number of preliminary facts have been grasped of which only a brief summary can be given here. For further information the reader is referred to the various text-books on wireless.

At the outset let us bear in mind that the basis of all these phenomena is induction, discovered by Faraday and universally applied in electricity at the present time. The following is a brief summary of the main features of this phenomenon :

An instantaneous electric current is generated in a conducting circuit whenever the magnetic flux which flows through it varies. The electromotive force of this induced current is all the greater, other things being equal, as the variation of the flux is *more rapid.* The phenomenon of induction has given rise to the theory of alternating current and to all the applications derived from it, notably to the use of self-inductance coils,

The Oscillating Circuit. What is an oscillating circuit? We know that before a circuit can be the centre of electrical oscillations it is essential it should possess self-inductance (spiral or coil) and capacity (condenser). When these conditions are fulfilled an electric or magnetic shock acting on the circuit so constituted gives rise to a series of oscillations.

According to the circumstances in which this phenomenon occurs, and to the way in which the source of energy manifests itself, for there necessarily must be in the circuit or in its vicinity some source of energy, the resulting succession of oscillations thus generated may be repeated and maintained.

Explanatory Analogies concerning Electrical Oscillations. For readers who are not familiar with the phenomena involved in the production of oscillations in an electric circuit, we propose explaining, in a very elementary manner, how this occurs.

capacity, circuits of harmonic resonance, etc. We know that the phenomenon of resonance forms the basis of all electrical oscillations. A second point deserves attention: electrical oscillations are propagated through insulators better than through conductors because the former do not absorb them. An interrupted circuit, that is to say "open" from an electrical point of view, may thus be the centre of radio-electrical oscillations which are radiated through space in the form of electromagnetic waves. A radio-electrical wave propagating itself consists essentially of an electric field and a magnetic field which follow the variations of the particular wave both in time and space. The circulation of high-frequency oscillatory currents originates from insulating materials mainly by virtue of the extremely rapid vibration of these electrical movements and also owing to the phenomena of self-induction and capacity.

Rôle of Self Induction and Capacity. The phenomenon of self-induction is, as its name indicates, only a particular case of induction which manifests itself in the circuit that gives rise to it, creating a kind of auto-reaction.

Self-inductance or, more simply, inductance, is the part of an electric circuit in which the phenomenon of self-induction manifests itself. This latter is produced by a variable magnetic field. Self-induction comes into consideration when this circuit is traversed by a variable electric current or by an equally variable magnetic flux.

Self-inductance or, more simply, inductance, consists practically of one or several conducting spirals generally arranged in the form of coils. The induction flux formed by the spirals is axial.

A rectilinear conducting wire possesses self-inductance, due to a magnetic field created in its vicinity by any current flowing through it. The wire may be considered as a spiral of infinite diameter.

Capacity. When two conductors close to each other and separated by an insulator are raised to a certain potential difference, continuous or alternating, an accumulation of local electricity results on these two metallic armatures, due to the electric capacity of this system. Owing to the accumulation of electricity resulting under these conditions, the name of condenser has been given to the apparatus capable of producing this phenomenon.

We also know that an insulator, placed between two armatures, which

For the sake of the uninitiated let us first take two comparisons.

Let us imagine the pendulum of a clock. This is a system which may be started in two different ways according as the conditions are those associated with either one or the other of the following two cases.

1. Suppose that the mass of the pendulum, immersed in water, possesses a paddle to slow down its motion. If the pendulum is deviated from the vertical position and then released, it will slowly return, owing to the resistance of the water against the paddle, to the vertical position (Fig. 4).

2. Suppose now that the pendulum is suspended in the air and deprived of the paddle. It is expected that under the influence of an impulsion the pendulum will oscillate to and from the vertical position. Its motion thus becomes oscillatory and the frequency of oscillations is equal to the number of times that the pendulum passes through the vertical line in one second (Fig. 5).

If an external cause acts upon the pendulum with the same rhythm and in the same direction, its oscillations will continue without a stop. Thus we see that when there is

cannot be the centre of any conduction current similar to those flowing through the two conductors is, nevertheless, traversed by electric currents called convection currents.

The laws of electricity state that the current flowing through a condenser varies in intensity as the capacity of the condenser becomes greater, as the electric tension is raised, and as the frequency *of this tension itself becomes more marked.*

But it is important to observe that even if the tension and the capacity are very low, it is nevertheless possible to obtain a current of great intensity provided the frequency be very great.

For greater frequencies than a milliard, for example, the capacities brought into play are sometimes so weak that they may appear non-existent or negligible. They are capable, however, of letting high-frequency oscillations pass through the air between two armatures separated by several inches and forming a condenser.

For still higher frequencies a distance of several metres between the two conductors, always constitutes an appreciable capacity, and it is thus possible, thanks to high-frequency phenomena, to make a current flow through an " open " circuit. This is rendered possible because conduction currents, passing through electric conductors, close up again owing to aerial capacity in the form of convection currents.

Generally speaking, two single wires, placed close together, form capacity as they may be raised to different potentials. For the same reason the two ends of a single wire have capacity in relation to their extremities and to the external medium.

no resistance to displacement such a system produces mechanical oscillations.

Let us now consider two water vessels joined at the base by a long tube of small diameter, and let us raise one of the vessels. The level of the water in the first will fall while

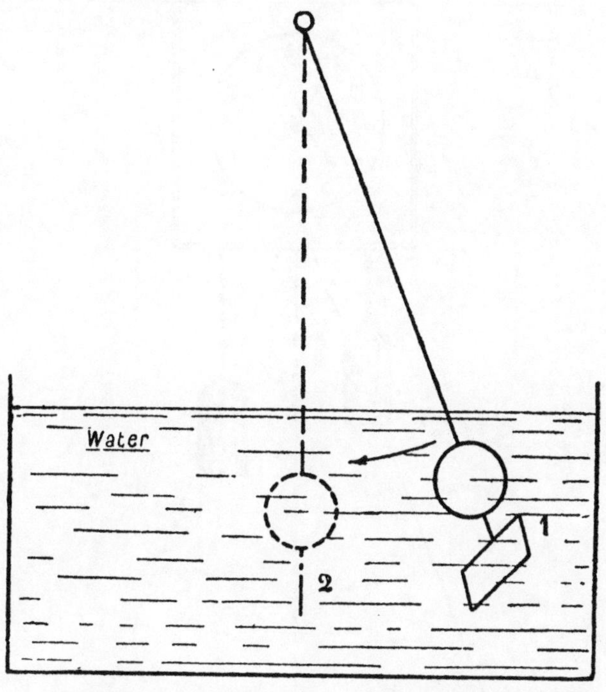

FIG. 4. *Motion of Pendulum in Water.* The pendulum being deviated from its position of equilibrium gradually resumes its original position without giving rise to any oscillations owing to the resistance of the liquid which damps down the motion.

in the other vessel it will gradually rise until the same level is reached in both vessels (Fig. 6). In this case, owing to the resistance of the tube due to its small diameter and great length, the final level is reached only by degrees in consequence of continuous displacement of water in the tube flowing in only one direction.

Let us now take a tube of short length and large diameter with a stopcock in the middle (Fig. 7). The stopcock being closed, let us raise one of the vessels to a certain height and then open the stopcock suddenly. We know that the final common level in the two vessels will be reached only after

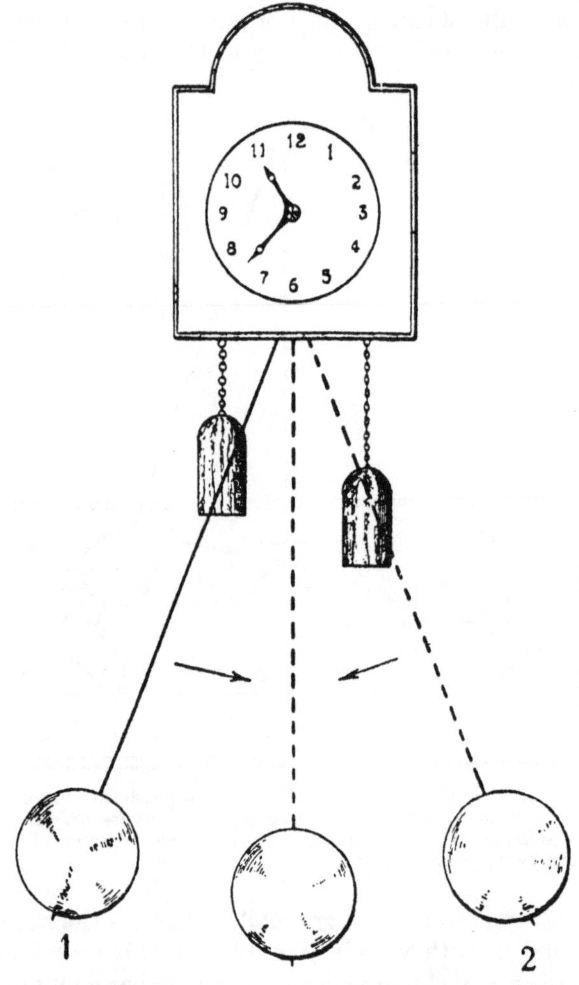

Fig. 5. *Oscillations of Pendulum.* The pendulum being deviated from the vertical to position 1, swings, by virtue of its own inertia, to a symmetrical position 2, and then swings back to the other side. It thus performs a series of oscillations whose motion is gradually damped down owing to friction of the axis of suspension and resistance of the air. It will ultimately stop and resume the vertical position.

The oscillations of the pendulum give a mechanical representation of electrical oscillations in a circuit consisting of self-inductance (inertia) and capacity (elasticity).

a few seconds, following a series of oscillations of the liquid contained in the respective vessels. This phenomenon of oscillation is due to the inertia of water as the result of the velocity acquired by the liquid and the sudden

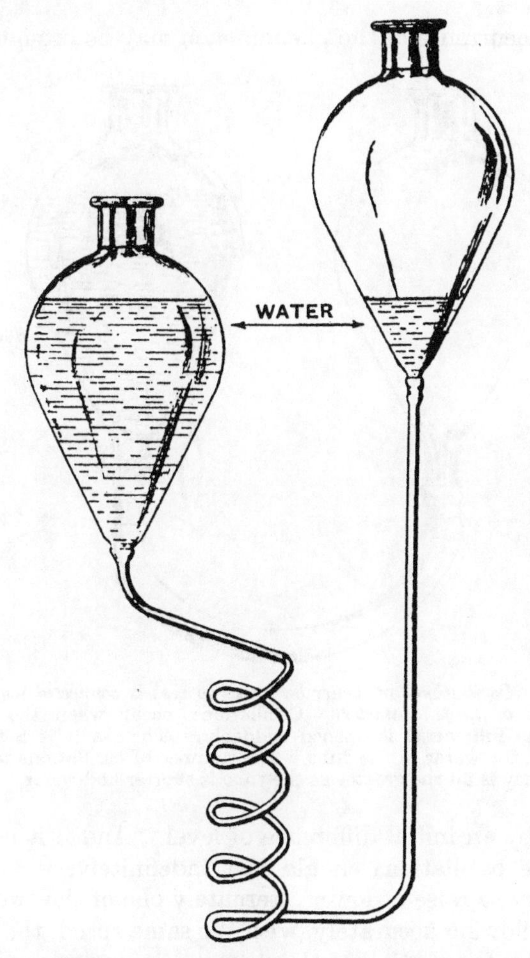

FIG. 6. *Oscillations of water between two vessels connected together by a long tube of small diameter.* In this case oscillations take place more slowly owing to the tube offering a high resistance to the displacement of water and also because more time is required by the water to travel from one vessel to the other.

If the resistance of the tube is sufficiently great the motion of the water will cease when equilibrium between the two levels is attained, and no oscillations occur.

motion it is subjected to in order to regain its position of equilibrium.

This state of equilibrium is reached only after a series of oscillations have taken place whose amplitude diminishes by degrees.

The occurrence of the phenomenon may be brought about

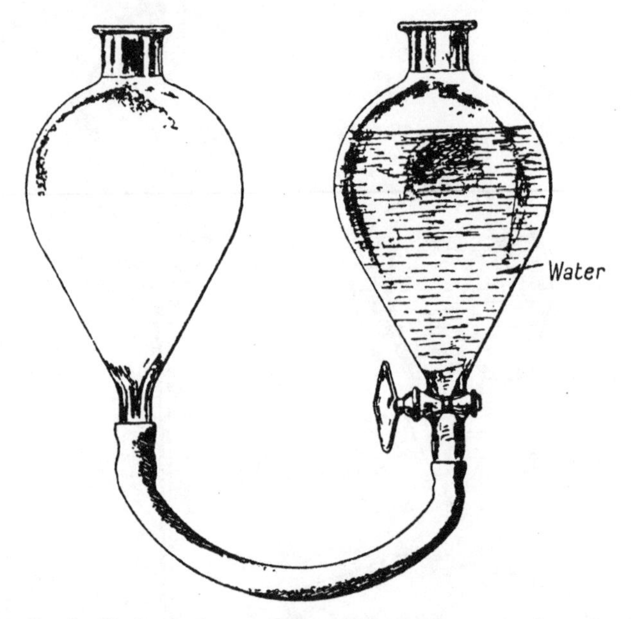

FIG. 7. *Oscillations of water between two vessels connected together by a short tube of large diameter.* Oscillations occur when the stopcock closing the full vessel is opened suddenly. The result is a to-and-fro motion of the water in the tube. The number of oscillations per second or frequency is all the greater as the tube is shorter and wider.

simply by an initial difference of level. And if it is desired that the oscillations should last indefinitely, it is merely necessary to raise or lower alternately one of the two vessels while following accurately, with the same speed, the rhythm caused by the motion of the water.

Thus we shall have produced, under the influence of an external cause, a permanent oscillatory motion of the water.

This simple and suggestive experiment is so familiar that we need not insist any further.

Let us note, however, three important points. The motion of the water is all the more rapid as :

1. The quantity of water is smaller.

2. The initial difference of level in the two vessels is greater.

3. The tube is less resistant, that is to say, bigger and shorter.

And now the same applies to electrical oscillations in an

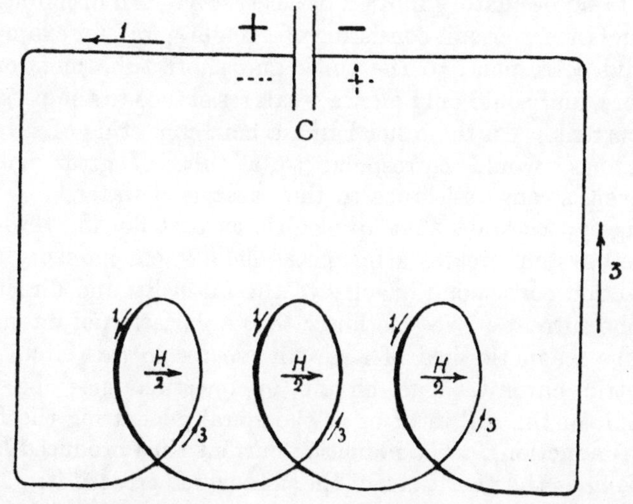

FIG. 8. *Theoretical explanation of Oscillatory Discharge of a condenser through self-inductance.* Above, the condenser is seen whose armatures are charged with positive (+) and negative (−) electricity respectively. Arrow 1 indicates the direction of the first discharge current. Arrow 2 indicates the direction of the instantaneous magnetic field H produced by this first current.

The production of this magnetic field H gives rise in the spirals, as the result of self-induction, to an instantaneous current whose direction is indicated by arrow 3.

It will be noticed that the direction is the same as that shown by arrow 1 and this current will duly charge the condenser. The condenser is thus charged with inverse polarities and is then discharged again, and so on. This is known as oscillatory discharge.

oscillating circuit formed, as we know, by self-inductance and capacity. The induction coil plays the rôle of the water vessel (Fig. 8).

The capacity of an electrical apparatus lies in its property of storing a quantity of electricity. The greater the capacity the greater its power of accumulating electricity. It is only required that the two metallic armatures of the capacity, separated by an insulator, should be raised to different electric tensions so that a charge may result. This capacity thus corresponds in every respect to the water vessel. But, instead of water charging the vessel, it is electricity that charges the capacity (condenser). Self-inductance corresponds to the volume of water contained in the tube joining the two vessels. The greater its action, the more it impedes the rapid oscillatory motion of electricity. An insignificant inductance, a circuit consisting of a single spiral, for example, would correspond to the thick and short tube mentioned before, and could only offer a weak resistance to the passage of current. On the other hand, a coil, consisting of several windings, would correspond to a tube of great length offering strong resistance to the passage of water.

Again, we know that an electric current flowing through a coil system creates a magnetic field whose intensity and direction correspond exactly to the intensity and direction of the current. We also know that a variation of intensity in the magnetic field of a circuit creates in this circuit an electric current. The circuit in question may be the circuit of the coil itself or of the spiral generating the field (self-induction). The induction current thus produced lasts as long as the variations of the field which created it.

To summarise: a current creates a magnetic field and the variation in a magnetic field gives rise to a variable electric current.

Let us further consider an oscillating circuit consisting of a spiral and a capacity formed by two metallic armatures separated by an insulator. Let us suppose that the circuit is open and the capacity charged. If the interrupter is closed, the capacity is discharged immediately into the spiral, giving rise to a current, even, as we observed before, in opening the stopcock, the water rushed into the tube. At the beginning the spiral is not affected by any current. Suddenly a current flows, rising from zero to a certain value. There is thus variation of current and creation of a variable magnetic field in the spiral, representing a certain variation of energy brought into play. But the current does not flow indefinitely and tends to fade out. The field created by the current will disappear and thus variation in the field will give rise, by induction in the coil system and the spiral, to an instantaneous electric current (see direction No. 3, Fig. 8).

Now, it is found, and it is a remarkable fact, that the direction of this induced current is the same as the direction of the first current of discharge, and that it tends to prolong its action.

It is the laws of induction which determine the direction of this current, and we shall not insist any further. But a new fact becomes already apparent. This current, supplementary to the primary current, charges in its turn the capacity which has just been discharged, only with an inverse polarity. All the energy of the discharge, which was transformed into electromagnetic energy, that is to say energy of motion, has been transformed again into electrostatic energy, that is to say potential energy, in order to charge the capacity in the inverse direction. But owing to various losses, notably through friction, which appears in the form of heat, this charge is smaller than the primary charge.

We now have a set of conditions similar to those at the beginning of the experiment : the condenser will be discharged afresh into the spiral, then recharged a third time with the identical polarity as the primary polarity.

The phenomenon will proceed on these lines until the complete exhaustion of the electric energy brought into play.

It will thus be seen that there will be a series of very rapid charges and discharges, that is what is termed an oscillatory discharge. This phenomenon comes to an end when all the energy is dissipated in the form of heat and radiation.

The rapidity of the succession of oscillations, that is to say their number per second, is known as the *frequency*. It is all the greater as the capacity takes less time to charge itself, that is to say as this capacity is weaker and also as the spiral is smaller.

It is easy to understand, therefore, the necessity of reducing as much as possible the spiral and the capacity in order to obtain very high frequencies. It is precisely what takes place within the living cells, as we shall see later. Moreover, we know that if the capacity and the spiral of an oscillating circuit diminish more and more, the wavelength may become as short as desired, but there is another thing which is reduced at the same time and very rapidly too, that is the energy brought into play. If the wavelength becomes extremely short the capacity will necessarily be very small and the energy almost negligible unless the

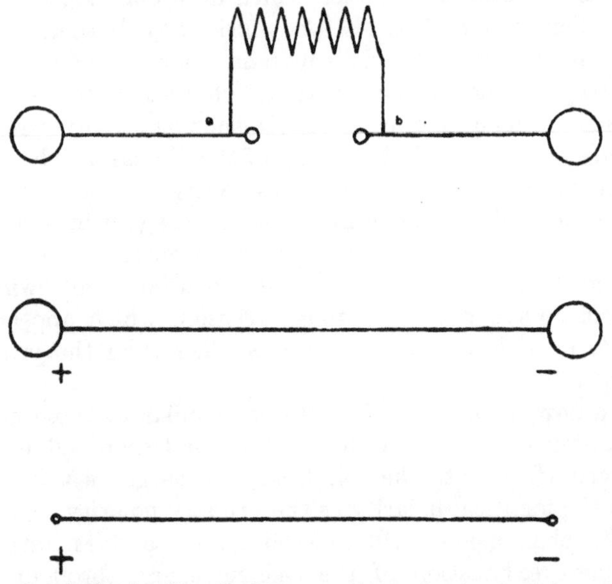

FIG. 9. *Oscillating Circuits of Hertz.* Above, the circuit of the oscillator of Hertz is seen. The secondary circuit consisting of an induction coil is connected with two balls or two metallic plates forming capacity by means of two wires, *a* and *b*, constituting self-inductance. An open oscillating circuit is thus obtained. The capacity formed by the two plates is discharged and gives rise to a spark between the two small balls.

The diagram in the middle shows a rectilinear oscillator consisting of a single wire (self-inductance) and terminating in two metallic plates or balls (capacity).

In the lower diagram the plates are reduced to the extremities of the metallic wire. The capacity is still existent, but it is very small. The frequency of oscillations is increased.

electric tensions employed are themselves considerable. But one is soon limited in this direction by the dielectric resistance of insulators and even by the air itself.

Let us recall to mind the experiments carried out by

Hertz with two metallic plates separated by a distance of 1 to 2 metres and raised to an alternating potential difference by means of a Ruhmkorff coil; the self-inductance was constituted simply by connecting wires and the condenser, by the capacity formed by the two plates suspended in the insulating air (Fig. 9).

This apparatus gives off wireless waves of short length. When the length of the connecting wires is diminished, as well as the diameter of the plates, the self-inductance and the capacity are equally diminished, but persist none the less.

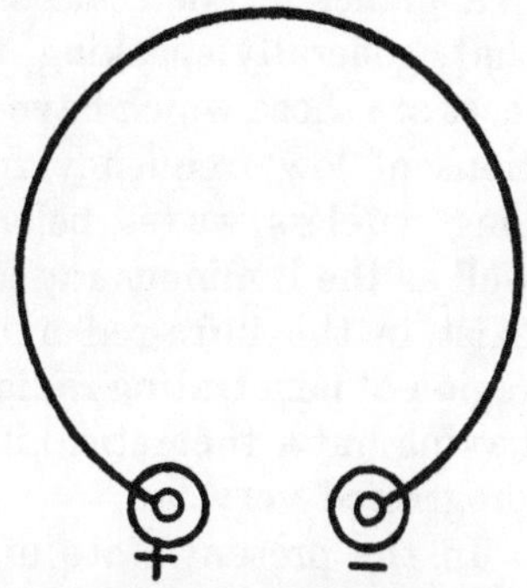

The apparatus may become microscopic, yet the oscillating circuit will always have a typical wavelength, but this wavelength will be correspondingly smaller and this also applies to the energy brought into play.

FIG. 10. *Schematic diagram of Electrical Oscillating Circuit showing similarity to Cellular Filaments.* This oscillating circuit may become microscopic. In the case of this diagram the extremities of the circuit are close together; they form capacity and take electrical charges, positive and negative. The small condenser thus formed is discharged into the wire forming self-inductance, in the same way as in an ordinary oscillating circuit. But the self-inductance is localised here along the filament.

Let us consider the particular case of a long rectilinear conducting wire whose two extremities are raised to any given potential difference. In relation to the material medium surrounding it, this wire is endowed with but a small degree of capacity and self-inductance. It can, therefore, be a source of electromagnetic oscillations of short wavelength, that is to say of high frequency.

The following three cases may be met with:

1. The circuit is subjected to any kind of electric or magnetic shock: it is then said that it vibrates according to its *natural period*.

2. The circuit is placed in a variable electromagnetic field or else it is subjected to the influence of electromagnetic

waves having the same frequency as its own frequency. It then vibrates, so to speak, in sympathy, or to put it more accurately, *in resonance.*

3. Under the influence of an external cause, the circuit may also be the centre of forced oscillations of a different kind of frequency. It is then said that it vibrates *aperiodically.*

A glance at the scale of electromagnetic waves will show that, generally speaking, the oscillations of which we know least are those which have the shortest wavelength. Oscillations of low frequency from alternating currents and the long wireless waves belong to the domain of industry, as well as the luminous radiations and X-rays. But there still exist in the infra-red and ultra-violet regions, and in the region of penetrating radiations, whole gamuts of frequencies having but a theoretical interest, the study of which has not progressed very far.

In the present state of our knowledge we may say that there is no definite break between the so-called electromagnetic waves, the calorific waves or infra-red waves, the luminous waves and the cosmic waves.

CHAPTER V

OSCILLATION AND RADIATION OF CELLS

Comparison of Cell to Oscillating Circuit—Constitution of Cellular Oscillating Circuit—Characteristics and Wavelengths of Cellular Radiation—Nature of Cellular Radiation.

Comparison of Living Cell to an Oscillating Circuit. In the light of experimental facts, both physical and biological, that have been discussed in the preceding chapters, we are now in a position to consider the basis of my theory concerning the radiation of living cells.

In the third chapter this first principle was enunciated: Every living being emits radiations.

From what we have just learned in connexion with our physical studies of electromagnetic waves, it follows that emission of radiations necessarily implies an oscillatory phenomenon. Furthermore, the most rudimentary living organism being constituted by a single cell, it seems evident that the simplest biological oscillation must be that which manifests itself within the cell.

We can thus enunciate this second principle, being more definite and proceeding naturally from the first:

Every living cell is essentially dependent on its nucleus which is the centre of oscillations and gives off radiations.

What are these radiations and whence comes the energy involved? Here are two questions I propose answering in the following pages.

Let us suppose that the geometrical dimensions of an oscillating circuit diminish gradually until they become invisible and microscopic. The spiral and the capacity of the circuit, which will also become microscopic, will still exist none the less. Thanks to these two indispensable factors, the circuit will continue to oscillate under the influence of causes which we shall examine later, and with a wavelength more and more reduced. This is precisely what takes place within the cells. Microscopic analysis reveals the presence of nuclei as shown in Figs. 10 and 11.

These nuclei are, as we shall demonstrate presently, actual electric circuits endowed with self-inductance and capacity and consequently capable of oscillating. These circuits oscillate according to a range of wavelengths whose magnitude depends essentially on the values of spirals and capacities. The waves given off are thus of electromagnetic origin, by virtue of the nature of the circuits, and are also

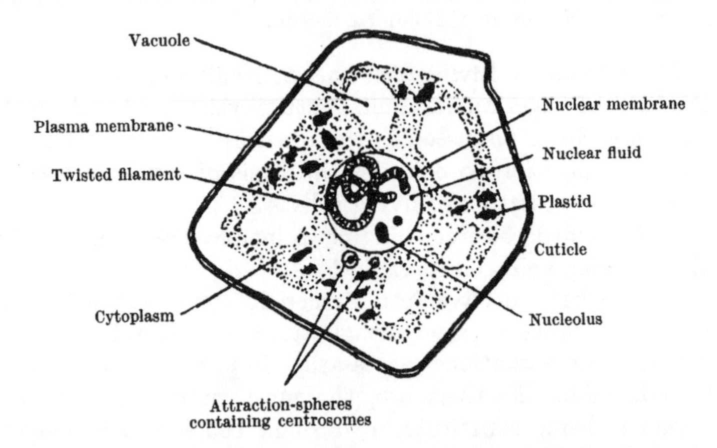

Fig. 11. *Microscopic view of various Elements entering into the composition of a Cell.* In the centre is the twisted filament which, possessing self-inductance and capacity, constitutes an oscillating circuit.

The similarity to a short-wave circuit is manifest: the filament shown here oscillates like a coil having a very small number of spirals.

of very high frequency owing to the minute dimensions of the organisms in question.

Constitution of Cellular Oscillating Circuit. Let us first call to mind what morphology teaches us on the subject of the constitution of cells. The details of cellular structure are made clear in Fig. 12.

A cell consists essentially of a nucleus or central system, immersed in protoplasm which is itself surrounded by a semi-permeable membrane. Examination of the nucleus reveals the existence of small twisted filaments constituting actual electric circuits. Fig. 12 shows a fragment of one of these filaments. They are composed of organic materials or mineral conductors, covered by a tubular membrane of insulating material consisting of cholesterol, plastin and other dielectric substances. Thus these organic structures, assuming the form of conducting filaments, constitute an electric circuit endowed by construction with self-inductance and capacity, which may well be compared to an oscillating circuit.

These circuits, characterised by extremely low values in regard to spiral and capacity, may under certain influences oscillate with a very high frequency and give off radiations of various wavelengths, just as the cells of the glow-worm give off visible radiations. The capacity and the spiral of

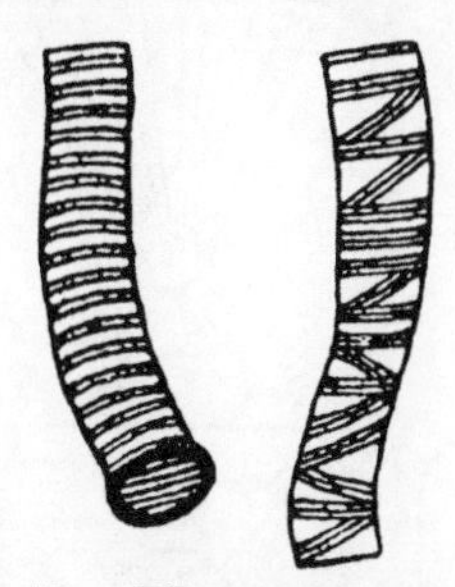

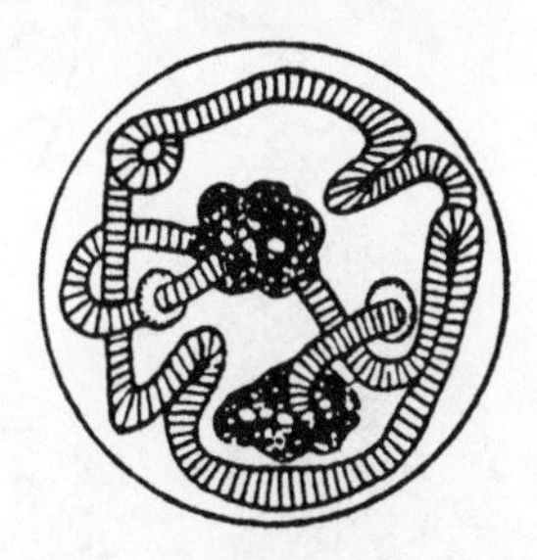

FIG. 12. *Filaments of Nucleus of a Cell.* On the left, fragments of filaments of cellular nucleus are seen. Their tubular structure should be noted. On the right is depicted a nucleus of salivary gland of the larva of *Chironomus plumosus* (after Balbiani).

these elementary circuits are, however, of a complex nature ; they depend chiefly on the form and the length of filaments, with their rings and sinuosities, together with the relative dimensions of the cell in regard to the filament. After a certain time and under the influence of a specific cause two mutually attractive poles arise in the protoplasm, the filaments are broken up, separated and orientated, to be finally united round each pole when the cell is then ready to divide (Fig. 13).

Characteristics and Wavelengths of Cellular Radiation. It is now clear, from the constitution of cells as revealed by the microscope and morphological studies, that each cell is capable of being the centre of oscillations of very high frequency, giving off invisible radiations belonging to a gamut close to that associated with light.

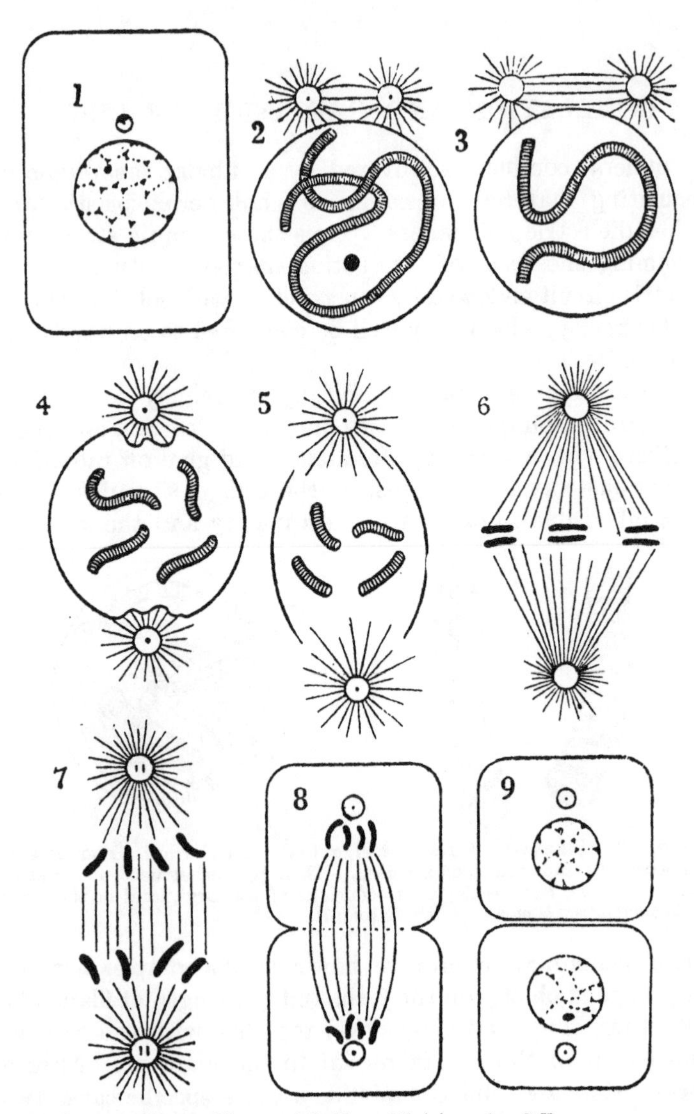

Fig.' 13. *Different Phases of Indirect Division of a Cell :*

1. Cell in the resting state with its nucleus and centrosome accompanied by attraction-sphere.

2. Isolated nucleus showing formation of filament ; division of attraction-sphere and outline of chromatin spindle.

3. Longitudinal division of filament.

4. Splitting of filament into four sets of chromosomes and depression of nucleus at the poles under the influence of asters.

5. The rays of asters penetrate into the nucleus and the membrane disappears at the poles.

6. Stage of " equatorial phase " : the chromosomes are orientated along a plane perpendicular to the spindle axis.

7. Separation of chromosomes which gravitate towards each attraction-sphere.

8. Cell whose cytoplasm begins to develop a " waist " in the middle, each half containing a nucleus in process of reconstitution.

9. Two daughter-cells resulting from division of original cell (after Henneguy).

Let us take, for example, the *Corynactis viridis*, magnified 1,000 times. From its actual size I calculated approximately the probable self-inductance of these intermingled circuits (Fig. 14). The capacity, however, is very difficult to determine. Taking certain average values, I found a radiation localised in the infra-red region. It is also possible to obtain a range of wavelengths—admittedly a rough approximation—by measuring the length of the filament and multiplying it by two. It is highly probable that the cells whose filaments are insulated at both extremities vibrate on the half-wave principle, that is to say have a wavelength nearly double the length of the filament, as the electric dipoles of Hertz. But these methods are not accurate and give but one type of wavelength. We shall see later why cells oscillate and under what influence. For the time being I hope I have convinced the reader that living cells are, according to their constitution, capable of oscillating and of emitting radiations.

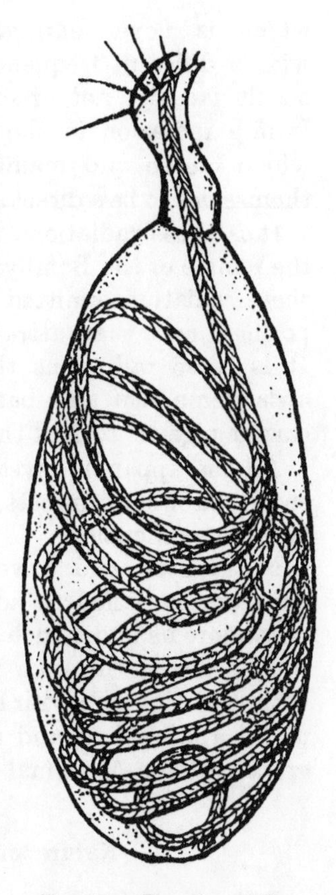

FIG. 14. Drawing of *Corynactis viridis* (magnification 1,000). In this marine organism, measuring but 0·1 mm., a number of internal circuits forming self-inductance, by virtue of the structural spirals, are clearly shown. Here the similarity to a self-induction coil is striking.

In the living organism the spirals may be seen drawing closer together or separating from one another. This results in modifications of wavelength while altering at the same time both the capacity and self-inductance of this remarkable variable circuit.

It is this phenomenon of radiation which lies at the root of the famous mysterious sense in birds and insects, that special instinct postulated by naturalists.

It is by means of this internal cellular radiation that the glow-worm produces its own light which is never extinguished. It is a similar radiation, with a different frequency, which endows insects with an occult faculty, not arising from the olfactory sense, but from a radiation in the ether. It is the same radiations which create and maintain life, or, at least, which show themselves to be a direct and inseparable manifestation of it.

It is these radiations that are emitted by the ovaries of the female of the Bombyx and that attract the males. It is these radiations emitted by the micro-organisms of decomposing meat that attract blue flies and burying-beetles. It is these radiations that direct, across great distances, owls, lemmings and bats towards their prey and enable carrier pigeons to find their course.

All the apparent mysteries involved in the instincts and social habits of insects, birds and other creatures, now become explicable.

Naturalists who have studied these phenomena have nevertheless failed to solve the problem that Nature has put before us under such a baffling aspect.

This theory throws a new light on the riddles associated with radiation and with life itself; it is susceptible of many useful applications and appears to be the keystone of the great problem of animal intelligence.

Nature of Cellular Radiation

It is gratifying to record that the investigations I have carried out in this field, largely inspired by the researches of Professor d'Arsonval and by the late Daniel Berthelot, have been confirmed by the recent experiments of Gurwitsch and Franck, as well as by those of Albert Nodon, President of the Société Astronomique of Bordeaux, who has been engaged for some years in the study of "actino-electric" phenomena produced in the living organism by ultra-short waves. These researches are particularly concerned with the radio-activity of plants and animals.

A. Nodon has carried out many experiments, with the aid of appropriate electrometers, with a view to comparing the radio-activity of plants and animals with that of mineral radio-active substances such as salts of radium and uranium.

The measurements recorded by Nodon were derived from many sources: leaves of hydrangea, pelargonium, leek, dahlia, ivy; grains of pollen, cloves of garlic, onion, potatoes freshly dug up.

It follows from these experiments that the so-called "radio-activity" is comparable to that of uranium, or to put it differently, that it causes the electrometer to discharge in 25–500 seconds, according to the nature and the mass of organic tissue. Extending his field of observations to animals, Nodon has shown that golden, black and green beetles, flies, spiders and other *living* insects, give off an amount of radio-activity equivalent to three to fifteen times the uranium value for an equal mass.

In passing, let us observe the fact, clearly confirming my theory of cellular oscillation, that dead plants and animals do not give any evidence of detectable radio-activity, for it appears that natural radiation is essential—and seems sufficient—for the maintenance of life. Indeed this radio-activity is but a manifestation of cellular oscillation. If the nucleus is destroyed oscillation ceases and the cell dies.

These observations, in addition to experiments on the human subject, have enabled Nodon to come to the following conclusion: "It appears from the recorded facts that the vital cells of the human body emit electrons generated by an actual radio-activity whose intensity would seem to be much more considerable than that observed in insects and plants." [1]

The fact that there should be a certain emission of energy in living beings, or a re-emission implying a previous activity, can hardly be doubted. The question is whether there is *transport* of energy by means of electrons or *transmission* of energy by means of waves. For my part I find it difficult to imagine that electrons may be transported over such considerable distances as those brought into play in certain biological phenomena, namely, instinct in animals and their powers of orientation, and the ways and means whereby their existence is maintained. There is every reason to believe that electrons are produced only locally as a result

[1] A. Nodon, "Les nouvelles radiations ultra-pénétrantes et la cellule vivante" (*Revue Scientifique*, October 22nd, 1927, t. lxv, p. 609).

of electric polarisation of organic tissues, but we must also bear in mind the actual phenomena of induction and detection in which waves play a leading part in the human organism, as the result of oscillation of an organic circuit consisting of the cellular nucleus.

Moreover, Nodon has obtained what may be called "spontaneous radiographs" by placing living things (plants, insects) directly on photographic plates. Clear pictures were duly registered after an exposure of several hours. Nodon's conclusion was as follows: "It seems probable that matter, under the influence of radiations whose wavelength is less than that of the diameter of the electron, may be subjected to certain modifications of unknown nature (?) which may confer new properties on matter, different from those conferred by radiations of much greater wavelength, and not connected with electrons."

The interpretation of these results appears to me to be much simpler. We are actually living in the midst of fields of cosmic radiations, comprising the whole range of waves, from the longest to the shortest. It must be obvious, as I have shown in the preceding chapters, that cosmic radiation induces in the cellular nuclei of the organism certain electrical phenomena, and, conversely, that the internal phenomena of the organism, notably nutrition, bring into play a series of electrical oscillations within the cells.

The theory which I have formulated on the oscillation of living beings accounts for these phenomena. The living cell is an actual oscillator and an electric resonator. Its "constants" are fixed by the form and the nature of substances entering into its composition. The renewal of these substances by means of nutrition gives rise to local electronic effects, due to electrons liberated by chemical reactions of the living organism, which modify the electric constants of the cellular nucleus, On the other hand, radiations emitted by living beings do not entirely consist of radio-active radiations, for there are also calorific, infra-red and luminous radiations (glow-worm, mushrooms, micro-organisms and animalculæ).

In this connexion let us mention the discovery made by Gurwitsch and Franck of the "mitogenetic rays" which are given off the stalks and roots of freshly cut vegetables, so long as the cellular nucleus is not destroyed. These rays have been identified as being similar in nature to ultra-violet radiations and their discovery constitutes an important confirmation of my theory of cellular oscillation.

At a time when the adherents of the emission theory of light are again confronted with the opponents supporting the undulation theory, it may not seem inopportune to reconcile the Newtonians with the followers of Huyghens by showing, as de Broglie has done, that the electron is, after all, but a system of waves. Therefore it is conceivable that cosmic radiations may integrate or disintegrate electrons within the atom. Again, the existence of more and more penetrating cosmic rays is being demonstrated frequently, and, at the present time, there is no justification for anticipating a minimal limit to the magnitude of ultra-short waves. Up till now the study of the highest frequencies has been handicapped by instrumental imperfection. Hence there seems to be no valid reason for postulating a "living atom," as conceived by Nodon. Indeed it seems simpler to conclude that all living organisms, whether plants or animals, consist of electromagnetic systems normally in equilibrium under the influence of a field of cosmic radiations combined with internal radiations such as those conditioned by nutritional processes. Excessive or deficient amplitude of this radiation must involve oscillatory disequilibrium which is fatal to the organism. This state of affairs may be brought about simply by variations in the characteristics of radiations which modify the functional activity of the transmitter or cellular resonator.

Certain physicists and radio-electricians have objected that my theory contradicts the facts, because cosmic rays are so penetrating that they can go through a mass of lead 7 metres thick or more, and therefore cannot make the nucleus of the living cell oscillate, which constitutes in itself an oscillating circuit of far greater magnitude than is commensurate with the action of cosmic waves.

To this objection I may say that cosmic waves cover the whole range of wavelengths, even those measuring several thousand metres, a fact observed by radio-electricians in the reception of all frequencies resulting in "atmospherics." Furthermore, each group of cells possesses its own frequency with its characteristic vibrations, and each individual frequency may be identified in the vast gamut of cosmic waves.

Finally we shall ascertain later the consequences of my theory of cellular oscillation by observing the effects of modifications in cosmic radiation following interference resulting from

1. activity of sunspots.
2. secondary radiation of waves absorbed by the soil.
3. therapeutic application of oscillating circuits.

CHAPTER VI

MODIFICATIONS IN CELLS AND OSCILLATORY DISEQUILIBRIUM

Oscillatory Action of Microbes—Experiment demonstrating Electrical Properties of Microbes—Effect of Radiations—The Radio-cellulo-oscillator—Therapeutic Tests on "Experimental Cancer in Plants"—Lakhovsky's Theory in Relation to Pathology of Cancer—Significance of Temperature of Human Body—Fever and its Function.

Oscillatory Action of Microbes. The knowledge we have acquired concerning cellular radiation enables us to consider, under a new aspect, the problem of the pathological condition of cells which, as we have seen, function as minute living resonators.

I have pointed out that life—a phenomenon of oscillation in the cellular nucleus—is the outcome of radiation and is dependent upon it for its maintenance. We can easily understand that life, considered as a harmony of vibrations, may be modified or destroyed by any condition causing oscillatory disequilibrium, particularly by the radiations of certain microbes which overcome the radiations of weaker or less resistant cells.

It is essential that the amplitude of oscillation should have an adequate value so that the organism may be in a sound defensive state against the harmful radiations of certain microbes. The microbe, as a living organism, vibrating with a frequency lower or higher than that of the organic cell, causes, in the living being, an oscillatory disequilibrium. The sound cell which can no longer oscillate normally is then forced to modify the amplitude or the frequency of its own vibration which the microbe overcomes more or less completely by induction. As a result of being forced to vibrate under abnormal conditions the cell can no longer function normally ; it is, in fact, a diseased cell. In order that it may be restored to health it must be treated by means of a radiation of appropriate frequency which, in recharging the cell with the required energy, achieves the dual purpose of restoring it to health and to its original normal state.

The action of this auxiliary radiation neutralises and overcomes the detrimental action of the microbe.[1]

It cannot reasonably be held that what is valid in the case of organic cells in living beings is not also valid in the case of microbes which likewise consist of individual cells. The microbes, constituted by a cell with a nucleus, also emit radiations. Whenever these elementary forms of life come in contact with highly organised beings, the result is what may be termed a " war of radiations " between the microbes and the healthy cells.

The problem confronting us is somewhat analogous to the dilemma in which a rescuing individual finds himself when rushing to succour a friend in danger. He sees him faced by powerful aggressors but he dare not make use of his weapons for fear of injuring his friend struggling with his assailants in an inextricable scuffle.

Similarly, harmful microbes and healthy cells would be equally exposed to any electrical or radio-active agency that might be employed to counteract certain detrimental radiations. It is difficult to destroy the microbes without injuring the host. Indeed, since the time of Pasteur, the main object has always been to kill the microbes. This method has a great disadvantage for it destroys, besides the oscillation of the bacillus, the oscillation of the cell in contact with it.

Experience in the treatment of cancer and tuberculosis with radium, X-rays and ultra-violet rays, has shown the great difficulties involved in this form of therapy.

Experiments Demonstrating Electrical Properties of Microbes. It is perhaps to be expected that some people may express astonishment that an electrical theory of life and of

[1] The action of the microbe on the living cell may be reduced to the action of an oscillation on another oscillation. It is essentially comparable to the forced vibration, induced by a small heterodyne generator in a resonating circuit tuned up with the incoming oscillation. The action of this local generator falls into line with that of the radiation which is " in resonance." According to the value of its frequency and amplitude, this auxiliary vibration modifies and modulates, to a greater or lesser extent, the initial vibration which may be reinforced or more or less eliminated.

the living cell should be extended to microbes, for until now microbes have not been studied from an electrical point of view.

Let us refer to an experiment, carried out by biologists, which demonstrates that microbes are endowed with peculiar electrical properties that have remained hitherto unexplained.

The microbe of typhoid (*Bacillus typhosus*) and the *Bacillus coli* are extraordinarily alike (Figs. 15 and 16).

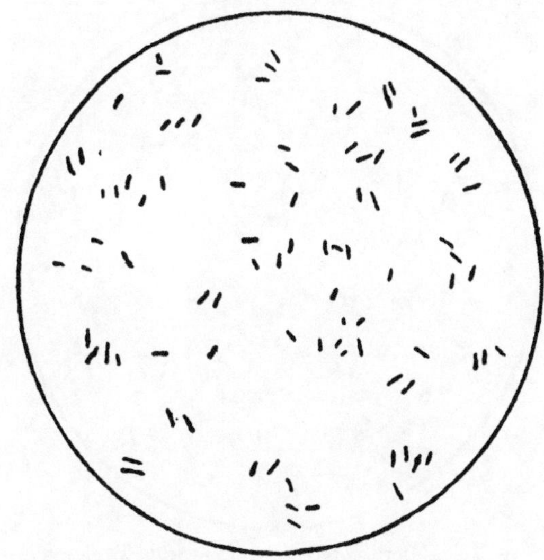

FIG. 15. Microscopic view of *Bacillus coli*.

The typhoid bacillus causes typhoid fever in man; it is found in the organs of typhoid patients and it can be cultivated. It is shaped in the form of a rod and measures 2 to 3 × 0·7 microns. This form of the bacillus may undergo modification. It is very motile, possesses vibratile cilia, and travels swiftly across the microscopic field.

As regards the *Bacillus coli*, it is invariably present in the intestine, in man as well as in animals. It is generally harmless, but it may become pathogenic. This latter variation resembles the typhoid bacillus, but it is less motile and shows but few cilia. It is also susceptible of cultivation.

These two micro-organisms were selected for the following experiment. A mixture of these two bacilli (*B. coli* and *B. typhosus*) was put in a liquid of slight electrical conductivity into which two electrodes were introduced and connected respectively with the positive and negative poles of an electric battery. It was then observed that the typhoid bacilli were attracted to one of the poles while the coli bacilli were attracted to the other pole. Thus the strict

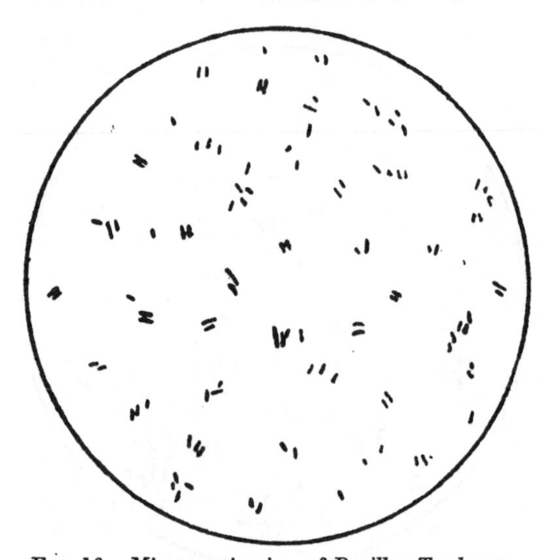

FIG. 16. Microscopic view of *Bacillus Typhosus*.

separation of the two types of bacilli was accomplished, the pathogenic and the non-pathogenic.

This experiment has even been filmed, and it is interesting to watch, as soon as the current operates, these microbes rushing, some to the right while others proceed to the left. This phenomenon, hitherto inexplicable, shows that microbes possess electrical properties of which we were not previously cognisant. Moreover, we know that in highly diluted solutions certain chemical compounds are dissociated, with the result that electrical charges appear, equal, but of opposite signs. For instance, sodium chloride, NaCl, is dissociated as sodium, Na, positively charged, and chlorine, Cl, negatively charged. Thus we may explain by analogy that the typhoid and the coli bacilli may undergo differentiation, from an electrical point of view, according to their chemical composition, just as sodium and chlorine are differentiated under suitable conditions.

In my opinion the coli bacillus becomes harmful only because it is capable of modifying, in general, the characteristics of the cell : capacity, self-inductance and conductivity. It follows that the coli bacillus, vibrating with the same frequency as the living cells, has no harmful effect on them, as it does not modify the wavelength of the cells. On the other hand, the typhoid bacillus, whose electrical properties are different, as the result of the differentiation of its chemical components, vibrates with another frequency, and modifies, by forced induction, the oscillatory equilibrium of the cell.

Effects of Radiations. With regard to the modifications effected by microbes in tissues and cells, let us endeavour, in the light of our theory, to find an appropriate remedy.

The problem is, not to aim at killing the microbes in the living organism, but to activate normal cellular oscillation by bringing a direct action to bear upon the cells by means of appropriate radiations.

My experiments have shown that with ultra-short wireless waves or with oscillating circuits in the form of collars and belts, it is possible to establish equilibrium of cellular oscillations and to overcome the effect of microbic oscillations.

The type of radiations produced by the waves in question is harmless, thus differing in that respect from X-rays and radium. Hence it should be borne in mind that their application is devoid of any risk. Furthermore, medical science makes use of high-frequency currents advocated by Professor d'Arsonval long before the discovery of the triode valve. This method has given excellent results.

The Radio-cellulo-oscillator. Having evolved a transmitting apparatus I experimented with a certain number of bacterial cultures which I subjected to the field of its influence for many hours. The result was that the cultures continued to grow normally. Moreover, I have never felt any malaise myself from these experiments although I was occupied for several days in manipulating this wave-generating apparatus to which I gave the name of *Radio-cellulo-oscillator*.

We are dealing here with an apparatus generating wireless waves, whose construction is immaterial, provided it produces the required radiation. The fundamental wave-length of this radiation is subject to variation. Its actual magnitude is conditioned by the nature of the cells undergoing treatment, but up to the present time I have used waves varying from 2 to 10 metres. It is only when living entities, such as the cell and the microbe, are in contact, that the rays given off by the radio-cellulo-oscillator are brought into action so that the oscillatory equilibrium of the cell may be re-established. It is the cell itself which, by recovering its vitality, thanks to the radiation of the auxiliary oscillator, succeeds in destroying the microbe.

The experiments which I carried out at the Salpêtrière Hospital with Professor Gosset, Dr. Gutmann and M. Magrou, were concerned with cancerous plants, inoculated according to the method of Erwin Smith. These experiments were the subject of a communication addressed to the Société de Biologie, on July 26th, 1924.

The text of this communication is given below.

Therapeutic Tests on "Experimental Cancer in Plants" [1]

Experiments have shown that it is possible to produce, in various plants, tumours comparable to cancer in animals by inoculation with Bacterium tumefaciens (Erwin F. Smith).[2] One of us [3] has obtained experimentally by this method a large number of tumours. These tumours continue to develop indefinitely ; under certain conditions they may undergo partial necrosis, but they do not perish entirely until the whole plant, or at least the branch bearing the tumour, succumbs to cachexia. Even when surgically removed these tumours invariably recur.

We propose to describe in this note the action of high frequency electromagnetic waves, generated by an apparatus

[1] By A. Gosset, A. Gutmann, G. Lakhovsky and J. Magrou.

[2] Erwin F. Smith, " An Introduction to Bacterial Diseases of Plants." Philadelphia and London, 1920.

[3] J. Magrou, *Revue de Pathologie comparée*, Mars, 1924.

PLATE I

PHOTOGRAPH SHOWING SCAR OF TREATED PLANT.

Pelargonium zonatum (Geranium) inoculated with *Bacterium tumefaciens* on April 10th, 1924, and treated from May 24th to June 14th, 1924, in eleven séances of three hours' duration, by means of Lakhovsky's oscillator fitted with antennæ.

Plant photographed after cure, July 21st, 1924. (*Surgical Clinic of Salpêtrière, Paris.*)

designed by Lakhovsky for therapeutic purposes, in accordance with his theories.[1] This apparatus has been named the Radio-cellulo-oscillator and gives off oscillations whose wavelength ($\lambda = 2$ metres approx.) corresponds to 150 million vibrations per second.

The first experiment began with a plant (*Pelargonium zonatum*) or Geranium taken a month after inoculation with Bacterium tumefaciens. It was affected at that time by small white tumours of the size of a cherry stone. The plant was exposed to radiation on two occasions at twenty-four hours' interval, and during three hours each time (Plate I).

For a few days following treatment the tumour continued to grow rapidly, like the control tumours, forming a great multilobar mass. About sixteen days after the first treatment the tumour suddenly began to undergo necrosis. Some time later (about fifteen days) the necrosis was complete; the lobes of the tumour, shrunk and desiccated, were separated by furrows of elimination from the stalk that bore them, and the tumour itself offered no resistance to the slightest traction. The necrosing action of the radiations was rigorously selective and strictly limited to the cancerous tissues which were attacked as far as the deepest site from which the tumours originated. The healthy parts, stalk and leaves, were left intact and the plant retained all its vigour.

A second geranium was similarly treated. In this case the duration of the exposure to radiation was prolonged (eleven séances of three hours each). Sixteen days after the first séance the tumour that the plant bore began to undergo necrosis and a few days later it was completely dried up. As in the first experiment, the healthy parts remained intact.

In a third geranium subjected to radiation during nine hours (in three séances of three hours each), necrosis of the lobes of the tumour followed the same course.

Sixteen geraniums were set aside as controls and were not treated. All of them bore tumours in full activity, often enormous (Plate II).

[1] Georges Lakhovsky, *Radio Revue*, Novembre, 1923, et Conférence à l'Ecole Supérieure des P.T.T. 2 Juin, 1924.

PLATE II

PHOTOGRAPH OF TUMOUR IN UNTREATED PLANT.

Pelargonium zonatum (Geranium) inoculated with *Bacterium tumefaciens* on April 10th, 1924, and photographed June 6th, 1924.

It will be observed that the stem of the plant bears a massive tumour. (*Surgical Clinic of Salpêtrière, Paris.*)

In conclusion, we are justified in stating that the geraniums that became cancerous after inoculation with Bacterium tumefaciens, a condition for which surgical intervention failed to prevent recurrence, appeared to be cured under the influence of certain electromagnetic waves previously mentioned in this communication. (Surgical Clinic of the Salpêtrière.)

The outcome of these experiments seems to be perfectly clear. On the one hand, a great number of plants inoculated with Bacterium tumefaciens and left untreated, have shown the development of tumours of considerable size that sapped their vital energy, ultimately causing their destruction. On the other hand, the plants treated by means of oscillations, and selected at random among the inoculated geraniums, were not only rapidly cured but were still flourishing, even in winter, while the geraniums not inoculated, duly produced flowers, but less conspicuously developed.

The remarkable photograph of a cured geranium on p. 89 (Plate III) should be of great interest to flower gardeners.

My Theory in Relation to the Pathology of Cancer

Statistics show that, in the majority of cases, cancer attacks middle-aged people, from fifty onwards, as well as a great number of old people, that is to say, cancer occurs in old tissues.

Our task therefore is to find out what chemical changes take place in the blood or in the cells of elderly people, for, according to my theory, cancer occurs as a result of variations of cellular oscillations caused by modifications in the electrical capacity of cells.

By way of example let us consider the formation of globulins.[1]

Analysis of the blood of elderly people has shown the presence of globulins rich in iron and phosphorus, built up from débris of fibrin, leucocytes (white corpuscles) and erythrocytes (red corpuscles). According to the investiga-

[1] A general name for various proteins, including globulin, vitellin, serum-albumin, fibrinogen, myosin and globin. (Translator.)

PLATE III

PHOTOGRAPH OF TREATED PLANT AFTER CURE.

This geranium is the same as that in Plate I, treated by means of Lakhovsky's oscillator on May 24th, 1924, and cured on June 4th, 1924. It was photographed in July, 1925.

As will be seen, this plant is in excellent condition and in full bloom.

On the other hand, the sixteen inoculated control plants, which were left untreated, perished long ago.

tions of several workers such as Achard, Aynaud, Bizzozero, Eberth, Hayem and others, there appears in the blood, from the age of 40 to 50, a number of flat corpuscles in the form of pellets, named globulins. Professor Aynaud has shown that globulins contain various mineral substances, representing one-sixth of the dry weight. The ashes of globulins show, on analysis, constant proportions of phosphorus, iron, sulphur and calcium.[1] In addition to mineral substances, globulins contain organic compounds such as lecithin whose chemical composition is akin to that of cholesterol which is found in all tumours of the skin.

Professor Roffo, the eminent cancerologist, has shown that cholesterol is found in all malignant tumours of the skin, Now cholesterol, according to Roffo's experiments, is susceptible to the influence of solar radiations, particularly those in the ultra-violet region. In a recent study [2] Roffo has established the fact that when the food of experimental rats is mixed with irradiated cholesterol (produced by sunlight or ultra-violet rays) malignant tumours (sarcoma) develop in 55 per cent. of cases, while in rats fed on non-irradiated cholesterol no tumours were observed at all.

The transformation of cholesterol involves the production of hydrocarbons which, by virtue of their radio-activity, act on the chromosomes of the cell which are destroyed, owing to "interference radiation," with the result that only mitochondria are left. These organic units being considerably smaller and having a far greater frequency than the chromosomes, continue to oscillate and to develop while also acquiring a cellular membrane. Hence the formation of the neoplastic cell.

The action of globulins in the causation of cancer now becomes more apparent. They contain, on the one hand, phosphorus (phosphorescent radiation in the presence of hydrocarbons found in globulins), and, on the other hand, mineral substances such as iron, calcium, sulphur, which increase the conductivity of the cells.

As in Roffo's experiments in which hydrocarbons caused

[1] Aynaud, "Sur la composition chimique des globulins." *Comptes-rendus de la Société de Biologie.* Paris, lxxvi, p. 480–481.

[2] *Bulletin de l'Instituta experimental para el estudio y tratamienta del cancer.* Buenos Ayres, December, 1937, No. 47.

cancer in mice fed with irradiated cholesterol, so, too, an excess of globulins in the organism from the age of 40 onwards, which introduces in the tissues the same hydrocarbons as in irradiated cholesterol, plays a part in the causation of human cancer. Furthermore, the researches of several investigators have shown that globulins agglutinate very rapidly and have a tendency to unite with organic particles, forming a covering which by its presence disturbs the oscillation of normal cells, finally giving rise to cancer.

It has also been observed that the number of white and red corpuscles is markedly smaller in old people than in adults, and according to certain investigators, both types of corpuscles, red and white, are transformed into globulins. Hence in old people the chemical composition of the blood is not the same as in adults.

Thus my theory provides a basis for the explanation of the phenomenon of cancer whose primary cause is still unknown, whether it be microbic or not. We also know that cancer may be grafted into a healthy organism but that the graft does not always " take." In such a case the normal oscillation of healthy cells overcomes the oscillation of the neoplasm (cancer) which fails to develop. On the other hand, if the graft is implanted into a group of abnormal cells such as " beauty spots," it often " takes " because the abnormal cell has a different rate of oscillation from that of the normal cell.

From these experiments it was finally concluded that cancer was not contagious and therefore that it was not due to a microbe.

From my standpoint I compare the cancerous cell to a micro-organism, having a nucleus just like ordinary cells, but whose frequency of oscillation is different from that of healthy cells. The only harmful microbes are those that destroy or modify the normal oscillation by altering the electrical capacity of cells ; and as for harmless microbes, it may be assumed that they vibrate with the same frequency as healthy cells, or that their chemical composition corresponds to the electrical capacity and resistance of the cellular environment. It is known that the lactic acid bacillus, yeast, etc., are not harmful any more than the coli bacillus under normal conditions, for, having the same oscillation as the healthy cells, they do not modify their frequency and therefore the cells do not undergo any alteration in spite of the presence of these bacilli.

Thus, in ageing tissues, an increase in the quantity of molecules containing metals (iron, etc.), due to globulins or other substances capable of modifying the electric constants of the cells and the blood, affects the internal capacity and electric resistance of every nuclear circuit. The circuit formed by the organic filament no longer possesses the same electrical capacity, indispensable for its equilibrium, for its specific wavelength has been altered. It follows that the frequency of oscillation is no longer the same. It has been definitely modified and differs from the specific frequency of healthy cells.

On the other hand, the division of cells which takes place as a result of the increase of the metal-containing molecules derived from the accretion of globulins or other carcinogenic substances, serves to increase the electrical capacity of other cells which causes a disturbance of their oscillatory equilibrium. As soon as the natural frequency is modified and the oscillatory equilibrium disturbed, the healthy cells, instead of dividing normally by karyokinesis [indirect cell division, the common mode of reproduction of cells] divide into neoplastic (cancerous) cells which vibrate with a different frequency. These new cells then act by direct induction and forced vibration upon the other neighbouring cells, which they force to oscillate with the frequency characteristic of cancerous tumours, and thus transform them into cancerous cells. The alteration of the tissues spreads by degrees and results in the appearance of a cancerous tumour.

Thus the primary cause of this alteration would seem to be a change of frequency on the part of healthy cells owing to an increase of globulins too rich in iron and phosphorus in cells already weakened.

At the age of about 50 certain organs undergo chemical modifications. The capacity and wavelength of the cells are also modified and they begin to vibrate with a different frequency, as stated before, forcing cellular division to become neoplastic (cancerous). The increase of globulins and other carcinogenic substances occurring at a certain age, in modifying the frequency of healthy cells whose electrical capacity is altered in consequence, or even in abolishing completely their normal oscillation, causes not only cancer, but also most of the diseases of old age. For cancer is but one of the diseases of old age ; it proclaims the degeneration of the organism.

I am convinced that ultimately we shall succeed in finding ways and means of regulating the capacity and the wave-length of cells. When this object is achieved there is no reason why human life should not be prolonged far beyond its present span. We observe, however, that in spite of modern hygiene, the mortality from cancer remains enormous. In my opinion this is due to a reason which should be rather reassuring, I mean the progress of science, paradoxical as it may seem. As a matter of fact, the average expectation of life (*i.e.*, mean duration of life) which was about thirty-nine years in the last decade of the past century, has risen to the figure of fifty or even higher in certain countries, thanks to the progress of surgery and hygiene, which has prevented a large number of deaths formerly due to contagious or organic diseases.

Cancer, so often incurable, attacks people who have reached the early fifties. The increase in the average expectation of life affects all classes of the community so that more and more people reach the " cancer age " and succumb to the disease.

In view of the rapid progress made in medical science and biology, and as some of the causes of cancer have already been definitely established, we are justified in hoping that this terrible disease will soon be conquered.

Significance of Temperature of Human Body

Fever and its Function. In the light of my theory it is possible to explain the phenomenon of maintenance of constant temperature in the human body.

Let us first consider how temperature is kept constant. Food, absorbed and chemically transformed by digestion and other internal processes, reaches every cell after having been assimilated partly by the blood and the protoplasm respectively. Food materials thus give rise to *biomagnomobile* entities which constitute the elementary units of living organisms, as molecules and atoms make up chemical substances. Foodstuffs convey to these units all the chemical elements, metals, metalloids, besides conducting and insulating compounds necessary for building up the organic filament, its nucleus and membrane. The nucleus is made up of two distinct parts.

1. Inside the filament, a mineral substance capable of maintaining to a certain degree the conductivity of the filament.

2. Covering the filament, a membrane consisting of a dielectric substance intended to insulate the filament itself.

We know that any oscillation in an electric circuit, open or closed, gives off heat produced by the passage of current through the conducting or insulating parts of the circuit. In other words, it is the friction of current against the resistance of the circuit which causes this production of heat.

In every cell the filament, consisting of conducting materials more or less electrically resistant, becomes overheated by the passage of current. Thus the fact that the cells oscillate implies that they give off heat, produced by the degradation of electric energy arising from the chemical energy of foodstuffs, and also from the atmosphere (cosmic rays) as we shall see later.

Let us suppose now that owing to any pathogenic agent the electric resistance of the filament of the nucleus and that of its membrane are different ; the result is an abnormal liberation of heat with repercussions on neighbouring cells. This emission of heat reaches the membranes of these cells so that the temperature of the body gradually rises and causes fever.

It is perhaps possible to correlate these facts with the death of certain patients suffering from high fever.

We have seen that the circuit constituted by the organic filament can oscillate only—that is to say, the cell can live only if this circuit, as any other electric circuit, is insulated from the liquid in which it is immersed. In fact the membrane of the filament serves a similar function as the silk or gutta-percha covering electric wires.

What happens, then, if the temperature reaches 41° C. ? Simply this : the insulating and resinous membrane consisting of plastin [1] or such-like substance, surrounding the conducting filament, fuses at this high temperature owing to its extreme thinness and its general physical nature. The circuit is no longer insulated ; it is destroyed. The cells, therefore, can no longer be the sources of electrical oscillations, they can no longer live, and they die.

The resistance, more or less prolonged, of certain patients to this high temperature is due to the particular chemical constant of the membrane of the nuclear filament, and to its degree of fusibility.

Acting on this principle it is clear that many diseases could be cured by means of fever, maintained at a suitable temperature, so that fusion of the nucleus of the microbe might be effected and the microbe consequently destroyed.

Thus we know that the gonococcus does not resist a temperature of 40° C. and that it is destroyed by the fusion of its nucleus following a fever exceeding this temperature. Moreover, for some time past fever has ceased to be considered solely as a pathological manifestation, harmful and inevitable. Indeed, remarkable cures have been attributed to fever which would seem to belong to the domain of empiricism, but will doubtless form part of the science of to-morrow.[2] Hence it is not useless to study closely the causes and effects of fever, for its artificial induction and adequate regulation depend upon such knowledge. We shall see presently to what extent my theory of cellular oscillation makes it possible to attain this end.

[1] A phosphorised protein constituting one of the chief proteins of protoplasm. (Translator.)

[2] Since Lakhovsky wrote this, therapeutic fever has developed into an accepted method of treatment known as Pyretotherapy. Application of heat by this method or Pyrothermy consists in a general heating of the patient with waves of about 30 metres. Pyrothermy has been applied by many workers in cases of rheumatic and other diseases, including general paralysis of the insane. Its aim is the production of artificial fever. (Translator.)

In this connexion it is interesting to note the occurrence of fever provoked by vaccination, and we may also recall that as early as 1885 Professor Wagner von Jauregg, of Vienna, indicated the possibility of treating general paralysis of the insane by inoculation with malaria, the same method apparently having been used to cure Louis XI of epilepsy.

At a time when microbes had not yet been discovered, the curative effects of fever had already been observed. Dr. Auguste Marie, an eminent French psychiatrist, mentions in a recent study the following observations made by Esquirol in his first treatise, dated 1818.

" There are few chronic diseases that have not been cured by the occurrence of an unexpected fever. All our practitioners invariably bewail their inability to produce fever. . . . Several have tried to induce it. . . ."

Generally speaking, cures in mental cases following fever have been observed on several occasions, especially when caused by malaria or erysipelas.

It is, of course, a purely empirical procedure to inoculate a patient with a disease and risk the consequences with the intention of effecting a cure by means of the resulting fever.

In my view the mechanism of the cure in question is quite simple. I have already stated that the nucleus of every cell consists of a certain number of substances whose nature and proportions are variable. Some of these substances act as conductors (mineral salts), while others act as insulators (resins, fats, cholesterol, etc.). They are arranged in such a way that the nucleus is generally found in the form of a tube made up of insulating matter (filament) filled with conducting fluid. Such are the elements of a cellular oscillating circuit.

Now these insulating substances are all fusible at various specific temperatures depending upon their nature. The membrane of the nuclear filament is thus an insulating compound which fuses at a certain temperature varying for each particular microbe, this temperature depending essentially on the nature and the proportion of the constituent elements.

The maximum temperature that a cell can withstand, without being destroyed, is naturally related to the constitution of the nucleus since the cell dies when its nucleus has been fused. Moreover, each species of microbe is resistant until a certain degree of temperature is reached. Observations made by various workers prove that a certain number of microbic diseases may be satisfactorily treated by means of fever provided the resulting temperature and its duration are adequately assessed.

But how is this to be done ? By resorting to malarial inoculation or colloidal substances which may cause grave organic disturbances that give rise to fever by reaction ? But fever may be excessive and cause fusion of the patient's healthy cells and thereby death may ensue.

I have also shown that fever originated from a rise of body temperature normally kept constant by the electric resistance, in the cellular-oscillating circuit, to the passage of high frequency induction currents. A rise of temperature in the cellular oscillating circuit may be brought about in two ways :—

1. Externally, by excess of induction current, arising, for example, from excess of cosmic radiation.

2. Internally, by diminution of electric resistance of the cellular filament ; for example, from excess of conducting mineral substances.

This is confirmed by many observations made on feverish patients.

In cases of fever, a rise of temperature is invariably observed in the evening, at sunset, when the sudden reduction of atmospheric ionisation due to sunlight causes a great influx of cosmic waves as well as short wireless waves. On the other hand, a decrease of temperature in fever is observed in the morning, at sunrise, owing to the diurnal attenuation of cosmic waves consequent upon atmospheric ionisation through luminous rays which interfere with cosmic waves.

In the light of these observations I believe it is a perfectly sound procedure to induce attacks of curative fever, not by inoculating patients with dangerous diseases or by destroying a microbe through introducing another in the organism, but by resorting to rational electrical methods, for example, by making use of an ultra-short wave generator, such as the apparatus I have already described, in addition to using oscillating circuits and appropriate resonators. The relatively long waves used in diathermy have a frequency which is far to low to generate a sufficiently high temperature, nor do they enable us to gauge accurately the localisation of the thermal effect produced. With much shorter waves, however, ranging from 1·50 to 3 metres, it is possible to bring about far more intensive heating effects.

The construction of a type of ultra-short wave apparatus of high energy potential has already been achieved to such a point that operators manipulating it have shown symptoms of high fever. An apparatus of this kind might enable us to regulate the intensity of an appropriate fever by generating the necessary heat in adequate quantity so as to fuse the nucleus of the pathogenic microbe.

I am of opinion that such a method of treatment might free mankind from many diseases, especially syphilis, which is one of the gravest, for we know that the spirochæte, its causative organism, is fused at a temperature of 40° C. Unfortunately, certain other microbes are fused at a higher temperature than our cells could tolerate, notably in the case of the bacillus of tuberculosis. In such circumstances artificial induction of fever would be impracticable and therefore, attention must be concentrated on increasing by chemical means the fusibility of the nucleus of the microbes in question, or diminishing the fusibility of our own cells, which would then enable us to use the ultra-short wave generator with a certain measure of success.

Further Proof of Cellular Oscillation

Sterilisation of Water by Direct Contact of Microbes with Metals. In order to prove the validity of my theory of cellular oscillation, I recently carried out a series of investigations at the Pasteur Institute. As microbes or cells can live only by virtue of their high frequency oscillation, and bearing in mind the bactericidal action of metals, I concluded that, according to my theory, the following facts provided a basis for a rational explanation.

It is known that the frequency of an oscillating circuit is modified by contact with a metallic substance which, in some way, short-circuits it. From this I deduced that the same phenomenon should occur in the cellular oscillating circuit, that is to say by contact of metal with microbe.

The experiments carried out at the Pasteur Institute confirmed once more my theoretical views, and formed the subject of the following communication presented by Professor d'Arsonval to the " Académie des Sciences " on April 15th, 1929.

Microbiology. Sterilisation of Water and other Liquids by Means of Metallic Circuits in Direct Contact therewith. Note of Georges Lakhovsky, presented by Professor d'Arsonval. (Abridged.)

The bactericidal power of silver has been known for some considerable time. Desiring to test the action of metals on microbes, according to my theory of cellular oscillation, which states that the nucleus of every cell or microbe is comparable to a high frequency oscillating circuit, and knowing that the frequency of oscillation of any circuit is modified by contact with a metallic substance, I concluded that the bactericidal action of the metal was purely physical and due to alteration of oscillation of the nucleus in direct contact with the metal.

In collaboration with M. Sesari, of the Pasteur Institute, I began these experiments with silver.

I. **Bacillus Coli.** An emulsion of *B. coli*, containing 11,280 colonies = 1,128,000 per cubic cm., was used as a standard. The emulsion was then distributed as follows into three separate vessels.

A—Used as control.

B—Circuit 7 flat spirals (surface area = 119 cm.^{2}).

C—Circuit 9 round spirals (surface area = 72 cm.^{2}).

After a certain lapse of time the results were as follows :

		Number of Colonies found	
		After 18 hours.	After 25 hours.
Bacillus coli per cubic cm.	Circuit A .	—	43,680,000
	Circuit B .	171,200	0
	Circuit C .	73,600	0

II. The same results were obtained with the typhoid bacillus. In this case the sterilisation process was slightly more prolonged.

III. In order to verify that the results obtained were not due to a chemical, but to a physical action of the metal, we carried out the following experiment.

After having mixed the water sterilised in the previous experiment with the silver circuits (Circuits B and C), we placed this mixed sterilised fluid into three glasses, *a*, *b*, *c*, as follows :

a—Without further treatment.
b—Heated between 101°–115° C.
c—Filtered wtih Chamberland F.

These three liquids were then contaminated afresh with *B. coli*, but in the absence of the silver circuits.

The titration, 10^{-1} : cm^3, at the end of twenty-four hours, gave the following results :

Control glass	10^{-1}	946 colonies.
Glass *a*	10^{-1}	12 ,,
Glass *b*	10^{-1}	13 ,,
Glass *c*	10^{-1}	1,474 ,,

It will be observed that the liquids *a* and *b*, containing the *B. coli* destroyed by the previous treatment, had an immunising effect on the newly introduced emulsion of *B. coli*, while in the filtered water (Glass *c*) the microbes developed normally.

We repeated these experiments with a white metal known as *platonix* with the same results.

From a hygienic point of view the conclusion is that a new process is available for the sterilisation of water without boiling (which renders it unpalatable and deprives it of certain mineral salts) and without adding chemical substances which affect its purity to a certain degree, and, lastly, without using filters which are not always effective.

I also wish to draw attention to the fact that the metal loses its bactericidal power when its surface becomes covered with a thin layer, consisting of calcareous deposits and organic matter derived from the water, which separates it from the microbes. The same phenomenon occurs in batteries and accumulators by polarisation when the electrodes have to be cleaned and depolarised.

The importance of this method of destroying microbes lies in the fact that without resorting to heat or chemical agents, it is possible to preserve the chemical constant of the microbe unimpaired, and this may conceivably extend the field of vaccination, especially in regard to the oral mode of treatment.

CHAPTER VII

NATURE OF RADIANT ENERGY

What is Radiant Energy ?—Ionisation and Conductivity—Deep Radiation and Cosmic Waves—Universion—Solar Radiation and Photolysis.

In the preceding chapters I have shown how the sense of orientation in animals could be accounted for and how living cells were centres of radiation. I now propose to consider the origin of these radiations.

Bearing in mind the relationship existing between radiation of healthy cells and oscillatory disequilibrium occurring in diseased conditions, I set out to reinforce this cellular oscillation by means of my high frequency oscillator, thus producing an extensive range of ultra-short waves susceptible of interfering with cosmic waves and of absorbing any excess in their output.

The existence of these interference waves is of the utmost importance for it seems clear that only certain waves of a frequency comparable to that of waves emitted by the cells can have an influence on the radiations of the latter.

In developing my theory I was faced with the problem of the origin of the energy necessary for the production and maintenance of cellular oscillations. Is it a question of chemical energy produced in living beings by internal radiations ? Or is it an internal energy of physical, thermal or luminous nature ? It does not seem probable, *a priori*, that it is a question of internal energy, any more than the electric battery, the steam engine or the dynamo, possess an energy of their own. Is it then a question of energy of external origin ? In point of fact it is actually a question of external cosmic radiation which astrophysicists have described as penetrating rays or cosmic rays which we shall consider in due course.

In order to ascertain the origin of this energy, I devised

PLATE IV

Photograph of another Geranium treated by means of an open Metallic Circuit.

This plant, inoculated on December 4th, 1924, was encircled by an open metallic circuit of 30 cm. diameter kept in position by an ebonite rod. The photograph, taken two months after inoculation, that is to say at the end of January, 1925, shows the tumour developing together with the plant which does not appear to be affected by it, whereas the control plants, inoculated on the same date and shown here beside the treated plant, have all perished.

the following experiment, similar to former experiments when plants artificially inoculated with cancer were treated by means of high frequency electromagnetic radiations which absorbed any excess in the output of cosmic waves at their maximum intensity. In this experiment I purposely dispensed with the local source of energy, that is to say, the Oscillator.

I took a series of geraniums previously inoculated with cancer, and placed them in separate pots. A month later, when the tumours had developed, I took one of the plants at random which I surrounded with a circular spiral consisting of copper and measuring 30 cm. in diameter, its two extremities, not joined together, being fixed into an ebonite support.[1] I then let the experiment follow its natural course during several weeks (Plate IV). After a fortnight I examined the plants. I was astonished to find that all my geraniums or the stalks bearing the tumours, were dead and dried up with the exception of the geranium surrounded by the copper spiral, which has since grown to twice the height of the untreated healthy plants (Plates V and VI).

What conclusion may we deduce from these results? That the copper spiral must have picked up external radiations, atmospheric radiations, and that it created an electromagnetic field which absorbed any excess of cosmic waves in the same manner as the Oscillator in my previous experiments. The corollary of this conclusion is that the atmosphere must be permeated with radiations of all frequencies. Indeed, we know that the terrestrial atmosphere contains a vast number of electromagnetic oscillations of all wavelengths and intensities, owing to constant and innumerable electrical discharges. Furthermore, we know that all types of electromotors and most electrical appliances create in the atmosphere a whole field of permanent auxiliary waves.

Again, during the past few years wireless stations have sprung up to such an extent that there is no detectable gap in the gamut of these waves. In such circumstances it

[1] An oscillator of this kind has a fundamental wavelength of about 2 metres and picks up the oscillating energy of innumerable radiations in the atmosphere.

PLATE V

PHOTOGRAPH OF GERANIUM SHOWN IN PLATE IV, AFTER CURE.

The plant is still flourishing and shows considerable development. The tumour has been shed and it may be seen in the foreground of the flower vase. On the stem the scar is clearly visible.

PLATE VI

PHOTOGRAPH OF THE SAME GERANIUM AS IN PLATE V, TAKEN A FEW MONTHS LATER (JUNE, 1925).

The plant is now completely cured. It continues to grow and bloom normally. As for the control plants shown beside it, they are all dead.

follows that any oscillating circuit of any dimension and of any shape may find, in this vast field of waves, its own particular wave which will enable it to oscillate normally. It is now clear that in order to attain this end it is unnecessary to have recourse to a generator emitting local waves, such as the Radio-cellulo-oscillator, with which I treated the inoculated geraniums in the course of my first experiments.

The question that naturally thrusts itself before us now is how do the oscillating circuit and the Radio-cellulo-oscillator act on the cosmic waves ? As we shall see in due course, it is the cosmic waves which create and maintain life by making the cellular circuit oscillate. Similarly, all electromagnetic waves, light, heat, electrical discharges, X-rays, ultra-violet rays, radio-active rays, etc., possess the property of reacting upon one another and upon cosmic waves. Experience has taught us that the intensity of cosmic waves is not constant, but is maximal at night towards midnight and minimal towards mid-day, as the diurnal radiation of light diminishes their intensity. These variations are detrimental to the maintenance of the oscillatory equilibrium of cells and may give rise to disease and death.

Owing to the action of the Radio-cellulo-oscillator or simply of the oscillating circuit which picks up the radiating energy in the atmosphere, and owing to the electromagnetic field thus created, excess of cosmic waves is absorbed.

In the following pages we shall discuss the nature of cosmic waves and how they affect the conditions of living beings.

Ionisation and Conductivity. Let us first remark that the subject of cosmic radiation will be made clearer if we bear in mind the following well-known fact. If a gold leaf electroscope, thoroughly insulated and placed under an airtight glass container, is charged, it will be noticed, after a certain time, that a progressive discharge takes place. If the experimental conditions are kept constant this discharge is stabilised and the wastage stops. (In certain experiments at the end of four days.) On the other hand, if the air is charged or if a fresh supply of air is introduced, the wastage continues.[1]

It has also been observed that this wastage increases in proportion as the pressure increases.[2]

Many scientists have studied this phenomenon, notably Geitel, Wilson and Campbell. Their observations have led them to conclude that the air was rendered conductive owing to a special cause, this is what is known as the phenomenon of spontaneous ionisation.

In order to ascertain the causes of this ionisation scientists have investigated the influence of radio-active radiation emanating from the walls of the container and depending on the nature of these walls. In short, they have determined the nature and manifestations of all the influences involved and have observed the following phenomena.

The spontaneous ionisation of air placed in an airtight container (washed and polished) is not constant. It varies with the time of day and attains a maximum towards midnight.[3] This ionisation often shows sudden variations which seem inexplicable, and it takes place equally well during the day or night, in cities or in the country. Furthermore, spontaneous ionisation varies according to the electrostatic potential of the air.

Lastly, and this is still more remarkable, after diminishing slightly in intensity up to a height of about 500–700 metres above sea-level, the intensity increases more and more with the altitude. Spontaneous ionisation increases rapidly with altitude, thus at 5,000 metres it is seven times greater than at the earth's surface.

Penetrating Radiation. We are thus brought naturally to the point of conceiving the existence of an extra-terrestrial radiation, coming from the sun, for example, or else from

[1] It seems fairly obvious that the enclosed air in the field of the electroscope should become electrified. If the air is renewed the new atmosphere must become electrified in its turn to the detriment of the electric charge of the apparatus, which explains the observed wastage.

[2] It is clear that the insulating powers of the atmosphere must decrease as pressure increases. The mass of conducting material and the number of molecules enclosed in a given volume function in direct relation to the pressure.

[3] Variations of ionisation show a marked similarity to variations of intensity observed in the propagation of waves, and, conversely, to variations of natural electromagnetic phenomena, known as "atmospherics."

other sources. This radiation has been given the name of *penetrating radiation*.

Such a radiation plays a part in the progressive ionisation of the atmosphere. As we have already learned, the intensity of a cosmic field increases with the altitude. It is natural to assume that these two phenomena are intimately related and are due to the same cause. This hypothesis is confirmed by the existence of a conducting atmospheric layer, known as the Heaviside layer, and situated at a height of 80–100 kilometres above the earth's surface. This zone is familiar to all radio engineers.[1]

Whence comes this radiation, this energy ? Does it come from the sun, the immediate source of all energy on earth ? It seems probable. Does it come from other stars more or less distant ? It is quite possible. But, in any case, one fact is certain, this radiation exists.

Solar Radiation and Photolysis. We may go even further and say that the atmosphere in which we live is permeated with a multitude of vibrations, electrical oscillations, etc., of known or unknown origin, and essentially characterised by different frequencies.

We have already pointed out that sunlight forms but a very small part of the whole range of vibrations originating partly from the sun and partly from the stars and even the Milky Way. It is impossible to deny the influence of the stars in this connexion. The tides, occurring twice a day, by the combined action of the moon and the sun, show that the most extensive mechanical work taking place on the earth is of astral origin. Why then should not the earth receive, from distant stars and from the Milky Way in particular, radiations of very small amplitude, susceptible of producing infinitesimal effects ?

Nature is the scene of a host of phenomena, alleged to be inexistent or inexplicable owing to our limited powers of perception, but whose effects manifest themselves nevertheless. Thus I postulate the existence of a multitude of radiations of all frequencies emanating from interplanetary

[1] The Heaviside layer is now generally known as the Kennelly-Heaviside layer, and is said to be ionised by the sun's rays. It has been held to account for " fading " of wireless signals. (Translator.)

space and traversing our atmosphere unceasingly. To this conception I have given the name of *Universion*.

Some of these radiations, the luminous ones, transmit through their rays a certain amount of solar energy and give rise to a process of synthesis in plants in connexion with assimilation of chlorophyll. This phenomenon, which holds good for the whole vegetable kingdom, was termed photolysis by the eminent French scientist, Daniel Berthelot. Thus light would seem to play an important part in the lives of plants and animals alike. In the vegetable kingdom synthesis of organic matter is accomplished with simple elements and with the intervention of energy directly transmitted by solar radiations (light, heat, infra-red, ultra-violet and cosmic radiations) which bring about this metamorphosis.

Penetrating Radiation (Cosmic Rays) in Relation to Life. It is actually these radiations, of very high frequency, invisible and imperceptible to our senses, which were supposed to act, according to a *modus operandi* we shall discuss presently, on the metallic circuit mentioned in my experiments with cancerous geraniums. It is these radiations which were responsible, in the inoculated plants, for re-establishing oscillatory equilibrium between healthy and diseased cells. These radiations, which were instrumental in curing diseased plants, emanated in my first experiments from my Radio-cellulo-oscillator. In the course of subsequent experiments carried out with a metallic spiral, the process was simpler in so far as it was the cosmic rays, filtered by the spiral, which were brought into action, finally restoring the degenerating cells of the diseased geranium to healthy activity.

Thus the purpose of these radiations is to maintain, by resonance and interference, the natural vibration of healthy cells, and to re-establish the vibrations of unhealthy cells by eliminating the radiations of microbes, differing as they do in amplitude and frequency.

It is these radiations which maintain the vital activities of plants and animals.

Cosmic Rays and Universion. The hypothesis of penetrating radiation has been fully confirmed by many astrophysicists, principally in America. Penetrating radiation is now identified with "cosmic rays," these natural rays which reach us across immense distances and consisting of a vast gamut of frequencies.

The discovery of gamma rays in the atmosphere some years ago led to the assumption that they were due to an emanation of radium contained in the terrestrial crust. But since then, experiments carried out in a balloon by Göckel showed that this radiation was at least as intense at a height of 4,000 metres as at the earth's surface, instead of diminishing with increase of altitude. It has been established that this radiation is approximately eight times greater at a height of 9 kilometres than at ground level. In America, Millikan and Bowen obtained significant results at a height of 15 kilometres and also at a depth of 30 metres in Muir Lake below Mount Whitney, at an altitude of 3,540 metres. These investigators discovered that at a depth of 30 metres of water the intensity of radiation was still sufficient to discharge an electroscope to an appreciable degree. In estimating, at a depth of 7 metres of water, the resistance of atmospheric absorption above the lake, it was found that cosmic rays could penetrate more than 37 metres of water, equivalent to a thickness of 1·80 metres of lead relatively to the absorbing power of this metal. Thus these cosmic rays appeared to be 100 times more penetrating than the hardest X-rays. The American astrophysicists repeated their experiments at Arrowhead Lake, deeper than Muir Lake, and also at great heights. They found that cosmic rays did not come from any particular direction, but seemed to come from all parts of space.

These rays constitute a spectrum extending over an octave and their highest frequencies are nearly 2,000 times greater than those of X-rays. These radiations range, in the scale of electromagnetic waves, as far from X-rays as these are distant from luminous waves. But in striking the earth these rays are partially transformed into softer secondary rays which are less penetrating.

The researches carried out by Professor Millikan and Dr. Cameron, among others, have enabled them to measure the intensity of cosmic radiation in ions per square centimetre and per second at sea-level. The frequencies of cosmic radiations have so far been extended to 2 octaves of the electromagnetic spectrum. Astrophysicists have shown that these rays were still detectable after having penetrated through 53 metres of water and 4 metres of lead.

According to Professor Millikan the origin of ultra-penetrating radiation is due to the most varied molecular and atomic changes occurring throughout space. It is the reason why he has made use of the general term "cosmic radiation." Thus the interplanetary vacuum is but a fiction since it appears to be filled throughout by cosmic waves radiated by all the stars and asteroids, by nebulæ and even by the Milky Way.

From the numerous researches of astrophysicists it appears that the existence of a range of cosmic rays permeating all regions of space and even intersidereal regions, is positively established.

The inter-astral vacuum is an obsolete notion as we know that this vacuum shows evidence of considerable radiating energy, all the more intense as it is more distant from the atmosphere, and propagated in all directions throughout space. Moreover, this radiation traversing the ether of the physicists permeates all material bodies, even those of the greatest density, as we have just had occasion to observe. All the manifestations of energy on earth of which we have knowledge, directly or indirectly, are but emanations of these cosmic rays which constitute the only possible intersidereal vehicles. Let us also bear in mind that the presence of terrestrial elements, the concentration of matter and the appearance of life, both animate and inanimate, are but manifestations of these rays. Finally, the motion of the stars is maintained by the energy transmitted by these cosmic rays.

In view of all these facts the suggestion of universal power derived from this conception of cosmic rays, should not be associated with the notion of absolute vacuum as implied by the ether of physicists. I believe that this ether is not the negation of all matter but rather the synthesis of all radiating forces, and therefore I have given the name *Universion* to the universal plexus of all cosmic rays.

Universion is a conception of the infinitely great, symbolised by the boundless *universe* ; and of the infinitely small, the granule of electrified matter, symbolised by the *ion* which is a world in itself. The infinitely great of the universe is, in fact, nothing but the integration of infinitely small ions.

I have elaborated this conception of Universion in another work to which readers are referred.[1]

Universion is ubiquitous and all pervading. Every moment we have evidence of its presence, as effective as it is silent. The material universe and life itself are but unstable phenomena. A certain variation of the body's temperature is enough to put an end to life and dissociate matter, thus restoring ions and electrons into the flux of universion whence they are mobilised by cosmic rays for the creation of new material combinations and living organisms.

Dissociation under the influence of temperature, pressure, electrolysis, photolysis, chemical reactions, electromagnetic and radio-active, electrical and photo-electrical reactions, such are the proofs of the existence and ubiquity of universion.

Let us not lose sight of the fact that Universion is a medium that revolutionises established conceptions, a medium where disintegrated elements are consigned and transformed into electrical particles. These conceptions need not astonish us for they reveal nothing more, in the continuity of the universe, but degrees of condensation.

The study of electromagnetic phenomena has upset the old mechanistic conceptions on the constitution of matter. And now, the study of Universion and cosmic rays will extend the bounds of science and enable us to solve the most absorbing problems of life—including telepathy and transmission of thought.

[1] Georges Lakhovsky, " L'Universion." Gauthier-Villars. Paris, 1927.

CHAPTER VIII

SUNSPOTS AND COSMIC RADIATION IN RELATION TO HEALTH AND LIFE

FROM the earliest times the influence of the stars on human life has been recognised. When science was undeveloped these notions, essentially intuitive and empirical, gave birth to astrology. At the present time, in view of our scientific knowledge, it need hardly be stressed that all these beliefs and observations should be rigorously examined.

In the preceding chapter a new concept, which we named Universion, was discussed. This may be regarded as a kind of substratum in which cosmic waves of all frequencies are propagated in all directions. The cosmic waves emanate directly or indirectly from the stars and it is clear that since they come from multitudinous sources and penetrate everywhere, they must have a spontaneous influence on our living conditions as they have already been shown to have an effect in the domain of physical phenomena.

We must now proceed to investigate scientifically to what extent these cosmic waves affect our existence and the scope of their influence.

Before considering the general problem attention should be focussed on particular cases of cosmic radiations, such as those emanating from the sun and the moon which play a singular and preponderant part in relation to the earth.

It has been shown by a Belgian engineer, M. P. Vincent, that lunar radiation was responsible for interference phenomena in the course of transmission from wireless stations. It appears that every week the recurrence of the phases of the moon corresponds with maxima and minima of intensity in the reception of electromagnetic waves.[1]

We are apt to forget that the sun, besides giving off luminous, calorific and actinic rays, also gives off electric

[1] Georges Lakhovsky, " L'Universion," p. 127.

and magnetic waves, especially during the eruptive periods of its protuberances or sunspots. Let us bear in mind that these sunspots are nothing but volcanoes and that the crater of a single one of them may measure as much as 200,000 kilometres in diameter, or more than fifteen times the diameter of the earth.

In addition to light and heat, the sun sends us electromagnetic waves whose magnetic force affects the magnetism of the earth and causes deflections of the compass. The electric force of these waves also gives rise to terrestrial currents whose intensity is sometimes such that it becomes impossible to telegraph or to telephone. Magnetic storms and terrestrial currents cause grave perturbations in the field of electrical communications, wireless or otherwise. Furthermore, the phenomena of ionisation caused by cosmic radiations emanating from the sun have, as a direct consequence, a marked effect in impeding the propagation of waves round the earth's surface. This results in ionisation of the upper layers of the atmosphere which renders it conductive, refractive and reflective, giving rise to " atmospherics " so familiar to radio listeners.

Another important proof that the sun and stars give off radiations besides those associated with heat and light, is given by the phenomenon of aurora borealis which often accompanies magnetic storms. It is known that this is due to the fluorescence of the atmospheric upper strata brought about by cathode and X-rays which form part of the stream of cosmic rays emanating from sunspots.

Some astrophysicists have correlated the occurrence and intensity of sunspots with certain concomitant physical phenomena. They have observed that terrestrial cataclysms, tidal waves, and especially earthquakes, seem to be associated with sunspots, and that the presence of these sunspots, considered in relation to the earth in a periodic cycle of twenty-seven days or so, may be held to account for the occurrence of " lunations " of the sun.

The cause of these perturbations is attributable to interference of these solar waves with the normal field of cosmic waves which play the chief part in the scheme of interastral mechanics.

A graphic representation covering a period of years and indicating the variations of intensity in geophysical phenomena, in electrical phenomena (ionisation, conductivity of gases, aurora borealis), of magnetic phenomena (perturbations in the terrestrial magnetic field, electromagnetic phenomena, etc.) shows that the different curves exhibit a remarkable degree of parallelism and that these phenomena follow closely the variations affecting sunspots. According to these curves it is clear that the variations of these phenomena are periodical and that the cycle of their manifestations occurs about every eleven years.[1] Without enquiring into the cause of this periodicity we are led to the conclusion that cosmic radiations emanating from the sun cannot be confined in their effects to physical phenomena, such as electricity and electromagnetism. They must necessarily play a part in biological phenomena also which are intimately connected with physical phenomena.

The study of this question has resulted in many observations which have seldom been adequately interpreted. In the wake of physicists, meteorologists have made a certain contribution to our knowledge of sunspots. In 1651, Riccioli announced that a relation existed between the appearance of sunspots and the state of the sky. In 1801, Sir William Herschel confirmed this observation. The astrophysicist, Baxendall, showed, in 1887, how the average temperature on the earth's surface was connected with the number of sunspots per annum, a fact which was confirmed by other observers.

In Mauritius, Dr. Meldrum showed, in 1871, that in tropical regions the number of sunspots determines the number of cyclones. This observation, however, has only been confirmed in the tropics, where the maxima and minima of storms accompany with striking regularity the maxima and minima of sunspots.

[1] This is in striking agreement with a statement by Sir James Jeans in his work, "Through Space and Time." Writing on the subject of sunspots, Sir James Jeans said "A careful study of cross-sections of trees frequently shows that the rings change gradually in thickness in a cycle of eleven years which coincides exactly with the sunspot period. The thickest rings were formed in those years when sunspots were most plentiful and we see at once that abundance of sunspots goes with abundance of tree-growth and so with moist summers." (Translator.)

Tropical rains also appear to be associated with sunspots. Rainy years appear to coincide with maximal sunspot activities while drought years reflect activities of minimal order.

In tropical regions where, owing to the absence of clouds, the effects of the sun are more direct and easier to determine, W. Koppen, in 1873, showed that during the year preceding a minimum of sunspots, the thermometer was 0·41° C. above the average temperature while during the year preceding a maximum of sunspots, the thermometer was 0·32° C. under the average temperature. Blandford explained this by pointing out that the excess of thermal energy transmitted by the sun, causes excessive evaporation of the seas, hence the lowering of temperature. Moreux observed that this did not apply to great continental surfaces where the elevation of temperature invariably follows the appearance of sunspots. But all these meteorological laws are, owing to their nature, far less accurate than physical laws. Nevertheless, they constitute, in so far as the effects of solar radiation are concerned, a valuable indication. Moreover, the problem of sunspots is less concerned with the qualitative and morphological aspects of the spots than with the total solar activity which brings cosmic waves into play. Again, the periodicity of solar activity is not so simple as it might appear and cannot be expressed in the form of a pure sine-wave. A vast number of harmonics superimposed upon the fundamental wave indicate that the actual periodicity of the sun is affected by that of other stars generating cosmic waves. Numerous observations made in Madras and Washington in more than one hundred different observatories, have shown that outside the tropics, solar radiation causes two alternating periods of rain and drought in the course of about thirty-five years. Such examples could be multiplied indefinitely. A similar periodicity has been observed in the drift of icebergs and in the variation of level in lakes. In particular, the period of eleven and a half years is very apparent in the case of the Victoria and Albert Lakes in Equatorial Africa while a period of thirty-three years seems to apply to European lakes. Generally speaking, direct solar activity is shown in all these natural phenomena.

The domain of meteorology serves as a natural transitional link between physics and biology. It would seem rational, therefore, to investigate in what measure cosmic rays, which condition physical and meteorological phenomena, affect physiological phenomena. This idea seems to have occurred to certain scientists at a time when the tendency was to attribute all solar activity to sunspots and when cosmic rays were unknown.

Sir William Herschel wrote in 1801 : " It seems probable, in analysing the period between 1650 and 1713, and judging by the normal yields of wheat, that a scarcity of vegetation occurred whenever *the sun appeared to be free from spots*."

In 1901 Moreux observed that the yield of wheat in France and throughout the world generally followed roughly the variations of solar activity. He then proceeded to investigate the influence of this activity on human organisms. He expressed himself on this subject as follows :

" In my capacity of Professor in a college I had exceptional opportunities for making observations. Although not being a medical man, I could not help observing a recrudescence of rheumatic affections and neuralgia, coinciding not with sunspots but with the strongest magnetic deviations due to solar activity. Furthermore, the total number of punishments appeared to be a function of deviations of the magnetic needle which seemed to indicate a kind of abnormal nervous excitement on the part of students . . . and possibly of professors too, at times of solar activity. I deduced from this that a relation could conceivably exist between wars and the sun, and I published this curve of correlation on several occasions before and after the Great War."

For my part, I conceived the idea of establishing from my personal observations and those of astrophysicists, the laws to which the biological effects due to cosmic rays are subject, and particularly those effects resulting from solar activity.

In comparing the charts of solar activity from the Observatory of Meudon with the statistics of wine-growing districts in Burgundy and Beaujolais, I have been able to show a parallelism existing between these statistics and the charts in question, and I concluded that the remarkable vintage years coincided with the years of recrudescence in sunspots.

These observations formed the subject of an original paper entitled "The influence of astral waves on oscillation of living cells," which Professor d'Arsonval was kind enough to present on my behalf to the Académie des Sciences. This paper is reproduced below.

Influence of Astral Waves on Oscillation of Living Cells. (Communication by Georges Lakhovsky presented on March 28th, 1927, at the Académie des Sciences by Professor d'Arsonval.)

"In my work, 'L'Origine de la Vie,' which Professor d'Arsonval has done me the honour of presenting to the Académie des Sciences, I formulated my theory of the influence of penetrating rays (cosmic rays) on living beings. I showed, in fact, that the nucleus of every living cell, manifesting itself in the form of a tubular filament consisting of dielectric matter and filled with a conducting substance, is comparable to an oscillating circuit having self-inductance, capacity and electric resistance. Living cells can thus oscillate with very high frequencies under the influence of cosmic rays emitted by the stars.

I have attempted to prove the validity of my theory by studying the influence of astral radiation (sunspots, comets, interference of astral radiations, etc.) on living matter.

My observations were based on the curves of graphs drawn by the astrophysicists of the Meudon Observatory; these curves showing, since 1845, the activity of sunspots, the incidence of magnetic perturbations and of polar auroræ.

These three curves are remarkably parallel. I set myself the task of studying the correlation existing between these astral radiations on the one hand, and the development of vital activity in plants and animals on the other hand. As in the case of any given individual, periods of fatigue and disease alternate with periods of good health, so, too, with fruits and crops in general, there are, for every kind of product, years of good quality and years of poor quality.

With regard to wine, according to the documentation established by the Chambers of Commerce of Bordeaux and Burgundy, I have noted that the remarkable years

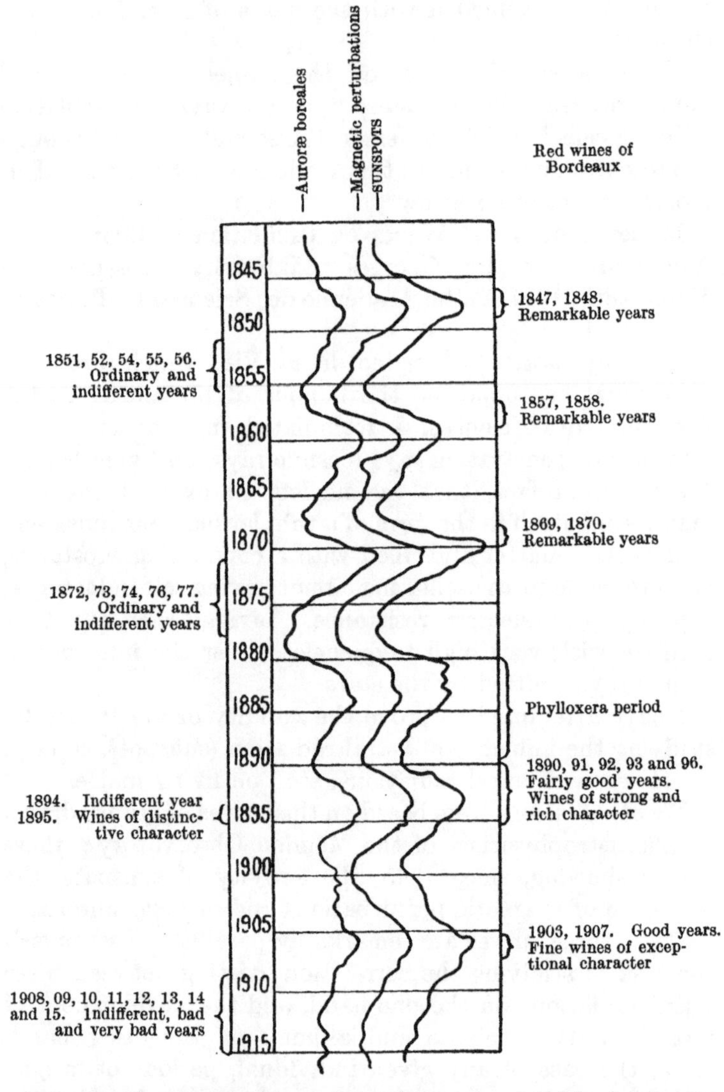

Fig. 17. Graph showing correlation between remarkable vintage years and intensity of solar radiations corresponding with variations of intensity in: (1) Sunspots, (3) Magnetic perturbations, (3) Auroræ boreales.

This graph refers to Bordeaux wines; maximal intensities correspond with good vintage years, while minimal intensities are associated with poor vintage years.

It is, of course, possible to draw up a similar graph for other wines, such as those of Burgundy, etc.

correspond exactly with a maximum of activity in sunspots, as the curves in Fig. 17 plainly indicate.

The results for red wines of Bordeaux are as follows :

Maximum, 1848 .	Remarkable years of 1847 and 1848
Maximum, 1858 .	,, ,, 1857 and 1858
Maximum, 1869 .	,, ,, 1869 and 1870
Period 1880–1889 .	Period of phylloxera.
Maximum, 1893 .	Fairly good years of 1890 to 1893.
Maximum, 1906 .	Good years of 1906 and 1907.

In this connexion special mention should be made of the famous wine of 1811 known as the ‘ wine of the comet ’—whose excellent quality may be attributed to the radiation of this comet.

The same results apply to the white wines of Bordeaux and Burgundy."

On somewhat similar lines a communication was addressed to the Académie de Médecine by Dr. Maurice Faure and Dr. G. Sardou.[1] These two physicians observed day by day and month by month the number of cases of sudden death and plotted a curve representing this phenomenon as a function of the weather. In comparing this curve with that representing the activity of solar energy they noted that these two curves showed a remarkable parallelism. Professor d'Arsonval remarked, in this connexion, that this appeared to be a particular case of my theory of oscillation in living beings.

It is not irrational to assume that interference brought about by sunspots may cause, if not disease, at least fatigue or transient disturbances. I have pointed out that periods of lassitude of the organism and of illness, and generally of disorders of sanitation might be attributable to interference phenomena which break up the oscillatory equilibrium of living cells. It has also occurred to me that these interference phenomena due to astral radiations, could provide an explanation of the modalities observed in the growth and development of living things in both the vegetable and animal kingdoms. It is possible that the flavour of a

[1] Académie de Medécine, session of March 1st, 1927.

certain fruit, for example, may be affected as a result of these interference phenomena. And if succeeding years differ from one another, from the point of view of agriculture, it is most probably due to variations of cosmic radiation. Thus we may account for good years, both in regard to quality and quantity, in the case of apples, plums, grapes, etc.

If I have stressed with some insistence the foregoing facts it is to show that although the question of the influence of solar radiation on the development of living organisms is not new, since the first observations were made over a century ago, yet it is only recently that the theory of cellular oscillation has enabled us to give an adequate explanation of it, thanks to our knowledge of interference phenomena.

It may be objected that the action of light and heat on plants and animals has been known for some considerable time. This is undoubtedly true, but light and heat are nothing but particular radiations of a restricted range in the whole scale of cosmic waves.

Evidence to the effect that light and heat do not constitute the whole output of solar activity may be found in the character of temperature curves, recorded in different observatories. These curves indicate that a multitude of local factors are involved, differing widely from one another, and, moreover, these curves are very unlike the curves representing solar activity in general. Furthermore, as we shall see in the next chapter, cosmic radiation is strongly influenced by the geological nature of the soil which, in its turn, may also give rise to interference phenomena.

In spite of their evident manifestations, light and heat have, at times, but secondary effects as compared with cosmic rays that remain imperceptible to our senses. It is possibly due to its elusive nature that cosmic radiation has hitherto passed unobserved, even though its effects are preponderant.

CHAPTER IX

INFLUENCE OF NATURE OF SOIL ON FIELD OF COSMIC WAVES

CONTRIBUTION TO THE CAUSATION OF CANCER

GEOLOGICAL AND GEOGRAPHICAL DISTRIBUTION OF CANCER

THE ROLE OF WATER IN RELATION TO CANCER

Nature of the Problem

THE studies in which I have been engaged for many years concerning the development and treatment of cancer have led me to investigate the causation of this disease which, at the present time, is the most mysterious and incurable affliction plaguing mankind.

I propose showing how my researches in this direction have led me to establish that the nature of the soil modifies the field of cosmic waves on the earth's surface. This condition may be sufficient to cause in living organisms a cellular disequilibrium susceptible of giving rise to cancer.

As no satisfactory evidence has yet been adduced in support of the contagious or hereditary nature of cancer, it seemed to me desirable to investigate the rôle played in the development of cancer by purely physical factors. Let it be clearly understood that by cancer or cancerosis, we mean the total number of cancerous affections, including carcinoma, epithelioma, sarcoma and other malignant tumours.

According to all medical accounts cancer is found in every part of the world, but the forms under which it appears vary in different regions. For some time past certain observers have assigned a particular rôle to different geographical factors such as orography and hydrography. In

1869, Haviland stated that " The Thames and its tributaries cover a vast cancer field." From the earliest times it has been observed that the morphology of living beings is closely connected with the nature of the soil upon which they live. The existence of different races adds support to this observation. Race is typified by marked physiological characteristics transmitted in a certain measure by heredity. But if living conditions are changed the characteristics of the race undergo transformation while still remaining bound to the nature of the soil and the climate. Several investigators have stressed the important rôle of the geological nature of the soil in the differentiation of racial types. The term " terroir " (smacking of the soil) which is used to describe the flavour of a certain wine, fruit or any product of the soil, implies clearly the preponderating influence of the soil in the elaboration of these products. Observations made in this connexion are very numerous and need not be mentioned here. Suffice it to say that plants grow indiscriminately on sandy soils as in the forest of Fontainebleau, but a strict selection occurs on clay and limestone.

As early as 1832 a pioneer naturalist, Nérée Boubée, informed the Académie des Sciences that the cholera epidemic which was then ravaging the country, was found to have a close relationship with the geological nature of the soil. Here is a characteristic passage from his communication: " In my annual geological travels I have often observed that in the countries where various endemic diseases occur, these diseases are most often confined, in every region, to the geological limits of the predominant formations, and I had already come to the conclusion that each geological region constitutes a natural stratum for certain morbid affections ; in other words, that the medical constitution of every country depends in some way on its geological and topographical constitution."

A few years later de Fourcault came to the same conclusions as Boubée in regard to other than endemic diseases.

Certain elementary considerations enable us to realise the influence of the geological nature of the soil and of its constituents. Water running through a certain region reflects exactly the chemical composition of the substances that constitute that region. In water are found the same mineral salts as in the soil. Again, the nature of water conditions the development of living organisms. In regions where calcium salts are deficient in the water the results are seen in deficient dentition and fragile bones. Let us also call to mind the influence of the nature of the soil in the causation of goitre, and generally, of hypertrophy or atrophy of glands resulting from excess or deficiency of a certain mineral substance in the soil of the habitat. It is, of course, well known that goitre which is a hypertrophy of the thyroid gland, occurs in regions deficient in iodine. Although the influence of the soil is indirect, it is none the less clearly evident. Nor can it be ignored that certain diseases exist in an endemic and latent state on certain soils where they remain localised. It is specially noticeable in the case of cholera, malaria and typhoid. Objections have been raised on the ground that these highly infectious diseases are transmitted only through microbes. It remains to explain, however, the reasons why certain microbes prefer certain soils, such as mosquitoes living on these soils. It is quite correct to state that cholera breaks out preferably on alluvial tracts while intermittent fevers are more commonly found on impermeable soils (clay or marl).

The influence of the soil is not only important in relation to pathological problems, but also in relation to hygiene and demography. Some time ago an Army doctor, M. Russo, sought to establish the influence of the soil on the health of the race. He showed that the most favourable conditions, from a hygienic point of view, occurred on soils of recent formation, tertiary or quaternary, followed by primary soils, granite and gneiss, jurassic, and cretaceous limestone.

In connexion with the cancer problem, M. Stélys, in a communication presented by Professor d'Arsonval to the Académie des Sciences [1] brought evidence in favour of *carcinogenic soils*, that is to say soils susceptible of giving rise to cancer in living organisms.

As the documentation concerning these various hypotheses and the co-ordination of the results obtained in this

[1] Session of April 25th, 1927.

field of investigation appeared to be sufficiently significant, I embodied the recorded data in a monograph entitled "Contribution to the Etiology of Cancer," which was presented by Professor d'Arsonval to the Académie des Sciences on July 4th, 1927. In this monograph I discussed the question of cosmic radiation in relation to the nature of the soil. Our present knowledge concerning cosmic waves and the propagation of ultra-short waves through different soils has proved an adequate basis to co-ordinate the various observations and statistical data. The object of this work was to show to what extent the distribution of cancer may be conditioned by the physical nature of the soil on which people live.

The problem of the etiology of cancer, considered from this point of view, has been conveniently reduced to the following three studies:

1. Demographical study of statistics on distribution of cancer, shown by the density of cancerosis or cancer mortality, calculated in number of cases per 1,000 inhabitants.
2. Geological study showing the soils on which cancerous tumours develop most freely.
3. Physical study, especially from the electrical point of view, of mineral substances constituting the soils in question and of the reactions of the latter to the penetration of cosmic waves.

Geological and Geographical Distribution of Cancer

The value of statistics in medicine has often been disputed, and it has been said that no reliance can be placed upon them. But statistics, however imperfect, constitute data that cannot be ignored. It is, at any rate, a definite indication that is preferable to the absence of any data at all.

Although the necessary ways and means of compiling statistics in villages and the country generally are lacking, this does not apply to urban areas, where exact information and abundance of data are available. Moreover, during the past decades it has been possible to diagnose cancer with a great deal of accuracy by means of microscopic and radiographic examinations which have made the classification of cancerous diseases possible. The number of actual errors inevitably involved in such statistics are thus reduced to a minimum and cannot invalidate the general tenor of the conclusions. Besides, all the investigations I have undertaken are based on statistics relating to cities and larger urban centres.

If the various districts of Paris are considered from the point of view of cancer density it will be seen at the first glance that the figures, far from being distributed in a haphazard manner, seem to vary in a *continuous* manner, in the algebraical sense of the term, that is to say without sudden solution of continuity. The same result appears clearly on the maps of parishes and towns. In these circumstances it is perfectly natural to think of a geological or geographical distribution of cancer.

The geographical distribution may be set aside without further consideration, for it would reduce itself to a mere survey of the land. The map of Paris, however, does not in any way establish the fact that the neighbourhood of the Seine or the factor of altitude play an important rôle in this connexion. On the other hand, the geological distribution gives suggestive results at the outset.

The problem we have to solve is why a relatively high cancer density affects the south-west and eastern districts of Paris, while the centre and north-west districts have a relatively low density.

Analysis shows that low cancer densities (0·5, 0·6, 0·8 per 1,000 inhabitants) coincide with a vast area of *sand and sandstone of Beauchamp* in proximity to limestone of the Paris basin. Medium, but still low figures, are observed in the districts of Chaussée d'Antin (0·8) and Gaillon (0·3), which correspond to an area of *sand of Beauchamp*. Higher figures, but still relatively low, are observed in Clignancourt (1·1) and Saint-Fargeau (1·04) where the only two outcrops of *sand of Fontainebleau* in Paris appear.

On the other hand, we observe that the districts where cancer density is high, such as Auteuil (1·76), Javel (1·61), Grenelle (2·08) and Saint-Lambert (1·57) rest on *plastic clay*. Other districts, such as Saint-Vincent-de-Paul (1·97), l'Hôpital Saint Louis (1·44), Père Lachaise (1·58) and

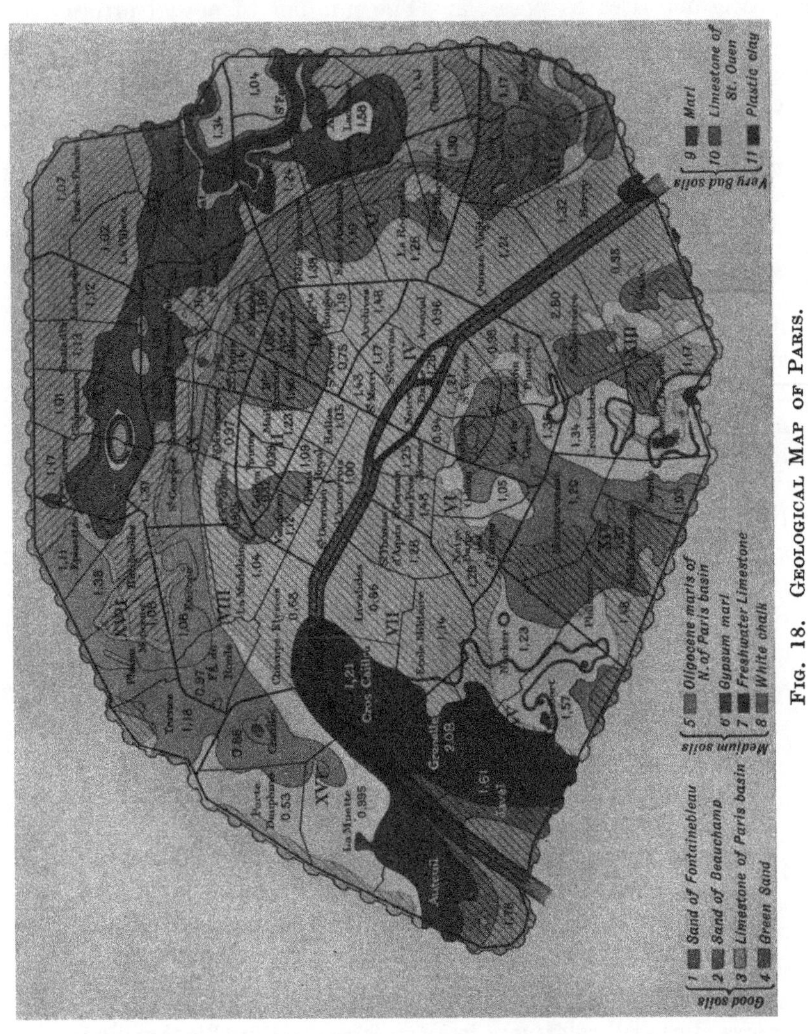

Fig. 18. Geological Map of Paris.
Influence of nature of soil on density of cancer incidence in Paris, expressed in numbers of cases per district and per thousand inhabitants.

Charonne (1·41) are situated on *marly soils* (Upper Oligocene of the Paris basin and Oligocene marls of the N. of the Paris basin).

The relationship observed between cancer density and the nature of the soil cannot be accepted as being mathematically correct as the geological distribution of the sub-soil presents a complexity as great as that of meteorological phenomena. Various perturbation factors must be taken into account, notably the disposition, the surface, the depth of strata and rocks, as well as variations in most sediments.

The district of Maison-Blanche, for example, with a medium density (1·17) contains in its sub-soil a mixture of clay, marl, limestone of Paris basin, sand of Beauchamp, and recent alluvial deposits. Similarly with Clignancourt (1·1) and Amérique (1·34), where we find sand of Fontainebleau, limestone of Brie and Saint-Ouen, marly and recent alluvial deposits.

With regard to the districts along the Seine, covered superficially by recent alluvial deposits, their cancer density reflects the composition of the deeper sedimentary layers. The same results are observed in the Département of the Seine in spite of the greater diversity in the nature of rocks.

Let us note that the localities with a low or medium cancer density, such as Sceaux (0·8), Chatenay (0·6), Bagneux (1), Fresnes (0·39), Suresnes (1·1), are built on *sand of Fontainebleau* while other localities, such as Garenne-Colombes (0·78), Vanves (1·18), Malakoff (0·98), Arcueil (1·27), Maisons-Alfort (1·29) are built on the *limestone of the Paris basin* or *the sandstone of Beauchamp ;* other localities, notably in the north-east of Paris are built on recent alluvial deposits and gypsum.

On the other hand, localities showing a high cancer density, such as Issy (2), Ivry (3·26), are built on *plastic clay ;* others, such as Les Lilas (1·63), Bagnolet (1·47), Pavillons-sous-les-Bois (1·91), Nogent (1·8), Romainville (1·85), Thiais (3·36) are built on *limestone of Brie* and *marl ;* finally, others, such as Neuilly (2·25), L'Ile-Saint-Denis (2·16), Le Perreux (1·87), Bonneuil (3·33), are built on muddy and clayey alluvial tracts.

I have applied the same method of analysis to the principal cities of France and those of neighbouring countries. The results have been grouped so as to indicate the density of cancer as a function of the geological nature of the soil.[1]

These investigations have clearly established the fact that

[1] For full particulars concerning this question, including charts, maps, etc., the reader is referred to Lakhovsky's special monograph on the subject, " Contribution à l'étiologie du cancer." Paris, 1927.

a low cancer density is found in localities built on sand, limestone, gypsum, sandstone, certain primitive rocks and recent alluvial deposits rich in gravel and sand. On the other hand, a high cancer density is associated with localities built on plastic clay, Jurassic marl, chalk, iron ores, carboniferous beds and slate.

It will also be seen that the cancer density in France is not distributed at random, but is related to natural regions corresponding to the geological nature of the soil.

Thus it appears that Geneva, Bern, Brussels, Antwerp and Toulouse are built in regions of medium or low cancer density, formed by sand and alluvial gravel, sand and sandstone of Fontainebleau and Beauchamp, limestone in proximity to marl.

On the other hand, the upper cretaceous formation covering the whole of Normandy, the Pays de Caux and Picardy, is noted for five localities having a high cancer density, Le Havre, Rouen, Amiens, Arras and Lille. Similarly, the east of France shows several regions with a high cancer density, characterised by iron ores (oolites, clay, ferruginous sandstone and marl) at Nancy and Metz, as well as carboniferous beds at Strassburg. The cancerous area of the Lyons region is also built on a Jurassic and carboniferous soil.

Nature of Soil in Relation to Cosmic Radiation and Causation of Cancer

The relation between the geological nature of the soil and the cancer density having been established by observations and statistics given in the preceding section, it remains to show by what particular mechanism a variation in the nature of the soil may bring about contributory factors in the causation of cancer.

I have already indicated, in a general manner, with regard to cellular oscillation, that cancer occurs as a reaction of the organism to a modification of its oscillatory equilibrium through the influence of cosmic radiations. Furthermore, the terrestrial field of cosmic waves is constantly affected by variations caused by interference phenomena due to various astral radiations, in consequence of the rotation of the earth either on its own axis (diurnal effect) or round the sun (annual effect) while the phases of the moon also affect the cosmic field.

Thus it is justifiable to establish a connexion between cancer and variations in the field of cosmic waves due to absorption by the soil.

We have seen that the oscillatory equilibrium of the cell is modified and sometimes broken up when cosmic radiations vary either in intensity or in frequency.

I have shown, however, that it was possible to re-establish this oscillatory equilibrium by reinforcing or diminishing, more accurately by "filtering" cosmic radiations by means of appropriate contrivances. Evidence of this was given by my first experiments on geraniums affected by cancer and successfully treated.

With regard to the absorption of cosmic waves by the soil and the resulting effects of these waves on the field, we have accurate data based on the labours of radio-electricians and astrophysicists who, like Millikan, have studied the problem of penetration. In this connexion, it is important to consider not only ultra-penetrating waves, but also the whole range of cosmic waves, from the longest to the shortest.

It has been questioned whether cosmic waves, in view of their great penetrability, have any effect whatever on the human organism. It should be borne in mind, however, that cosmic waves have such a universal field of action that it seems obvious, even *a priori*, that nothing can escape their influence. We also know that it is not necessary to stop the motion of a wave completely in order to detect its effects. At this rate, the detection of wireless waves would be possible only provided immense metallic walls of great thickness were available in order to capture the waves *in toto*. But all that is required to attain this end is a simple wire stretched out in the open space, which retains from the passage of the waves an unappreciable and yet sufficient amount of energy. Similarly, the living organism has no need to be like a mass of lead of 10 metres thickness in order to be sensitive to the induction of cosmic waves, to which it will respond most readily as the waves are of shortest length and the living cells of smallest dimensions. It is also clear that owing to the excessively high frequency of these cosmic waves, the cells must be subjected to a formidable electromagnetic induction.

Since we are able to detect, as Millikan has shown, cosmic waves at a depth of more than 50 metres, it is evidently not the total absorption that is of primary importance for, from a practical point of view, this is insignificant and must always depend on the sensitiveness of the apparatus employed. It is almost beyond doubt that certain cosmic waves exist which are sufficiently penetrating to traverse the whole earth, an hypothesis which seems to be essential to explain the phenomena of celestial mechanics.[1] What is of great importance, however, in investigating the influence of a certain phenomenon on the conditions of life, is to pay special attention to variations of the cosmic field at the earth's surface, which involves absorption by sedimentary layers and the resulting secondary radiation, as well as the interference field. This secondary radiation is no more negligible in the case of cosmic radiation than in that of radiological and ionisation tubes, which give off cathode rays and X-rays. In cities, the influence of building materials such as stone, bricks, masonry, tar, asphalt, paving-stones, need not be considered for these eminently dielectric materials do not impede the propagation of waves. We know that waves penetrate into the soil all the better as the insulating properties of the soil are more marked, which is in accordance with our knowledge of the propagation of waves. With a wavelength of 16,000 metres penetration is effected to a depth of 80 metres in an insulating soil (sand, limestone, etc.), whereas penetration reaches only a depth of 2 metres in sea-water which is a very good conductor ; and a few dozen metres in plastic clay and various ores, which are also very good conductors. The depth to which the wave penetrates into the soil is inversely proportional to the square root of the product of its vibration and the conductivity of the soil. Variations of penetration are thus much more marked in the case of short waves than long waves. The conducting soils act almost like metallic screens and

[1] **Georges Lakhovsky, " L'Universion." Paris, 1927.**

absorb waves to a maximum degree. On the other hand, the dielectric (insulating) soils facilitate the penetration of waves to a great depth. Thus it follows that these soils, permeable to waves, such as sand, sandstone and gravel, which absorb radiation to a great depth, do not show any appreciable reaction on the cosmic field at the earth's surface, as is the case whenever a wave penetrates a medium that is practically homogeneous and unlimited. But when the radiation is only superficially absorbed as in the case of

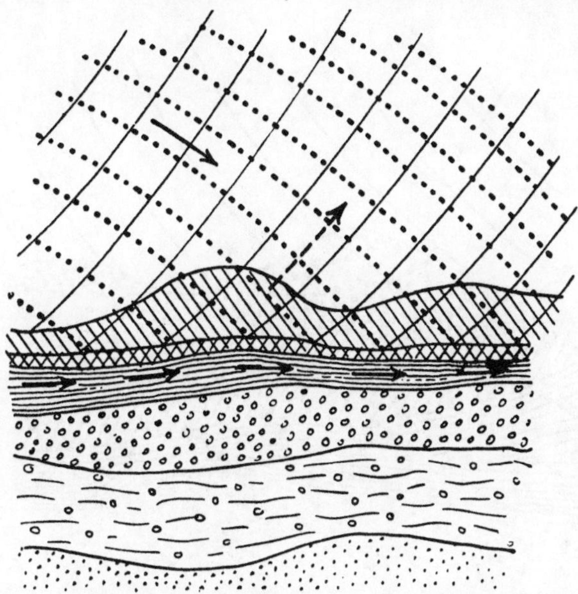

Fig. 19. *Conducting Soil Impermeable to Waves.* Cosmic radiations are reflected and diffused superficially, giving rise at the surface of the soil to a new field of interference radiations.

conducting soils impermeable to waves such as clay, marl, carboniferous beds, iron ores, this rapid absorption gives rise, at the surface of the conducting stratum, to intense currents which react on the superficial cosmic field.

It is possible that this absorption may give rise to refraction as is the rule in physics generally when the constants of the medium of propagation vary, for example, when luminous rays pass from air to water. Or else it may be that we are confronted with a more complex phenomenon in which absorption of cosmic radiation by the soil is followed by a secondary radiation or re-radiation. Be that as it may, it cannot be doubted that the secondary radiation, reflected, refracted or diffused by the conducting layer, interferes with the incident radiation, which results in a field of complex radiation different from the initial field (Fig. 19). On the other hand, in insulating soils cosmic radiation is not affected by the absence of secondary fields (Fig. 20).

As the development of cancer is supposed to be connected

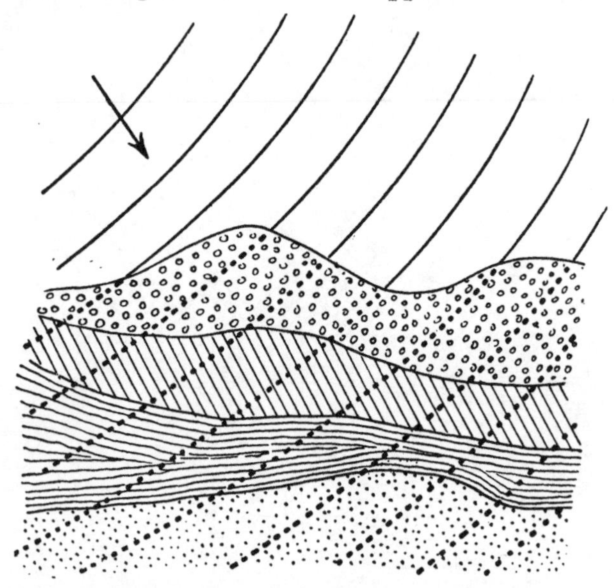

FIG. 20. *Insulating Soil permeable to Waves.* The superficial field of radiations is not modified. In this case there is no reflection of waves, no diffusion and no re-radiation.

with oscillatory disequilibrium caused by variations in the field of cosmic radiation, it follows that the incidence of cancer is low on insulating soils and high on conducting soils which modify the field.

The question of the influence of the soil on the incidence of cancer may thus be reduced to determining its degree of conductivity.

We have seen that a low incidence is found on the sand of Fontainebleau and on the sand of Beauchamp, which consist of pure silicates, and as such, are highly insulating; a low incidence is also observed on the sandstone of Beauchamp and on the sand of the Brussels basin, the gravel of Geneva and the friable sandstone of Bern ; the slate, gneiss and granite of Nantes ; the gypsum of the north-east of Paris.

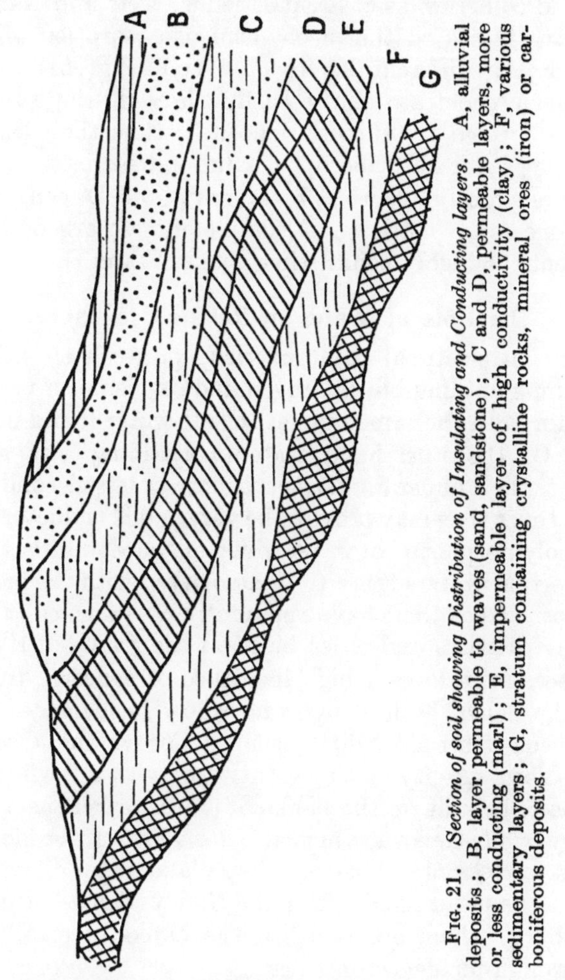

Fig. 21. *Section of soil showing Distribution of Insulating and Conducting layers.* A, alluvial deposits ; B, layer permeable to waves (sand, sandstone) ; C and D, permeable layers, more or less conducting (marl) ; E, impermeable layer of high conductivity (clay) ; F, various sedimentary layers ; G, stratum containing crystalline rocks, mineral ores (iron) or carboniferous deposits.

A medium and high incidence of cancer is found on soils which are fairly good conductors, such as recent alluvial deposits containing muddy beds of conducting soil, and especially plastic clay, by virtue of its chemical composition, including water and mineral substances.

The degree of cancer incidence increases on such soils as gypsum, marl (Upper Oligocene of the Paris basin) and Jurassic marl, impermeable clay, ferruginous limestone, ferruginous chalk. The incidence is highest on the soils containing ores and collieries, as at Saint-Etienne, Metz and Nancy.

I have indicated the mechanism of absorption of waves through the different layers of a soil (Fig. 21). Cosmic radiations penetrate fairly easily through the superficial layer A formed by alluvial deposits ; then they reach the insulating layer B, made up of sand and sandstone ; radiations are slightly absorbed by layers C and D, consisting of limestone and marl, and completely absorbed by the sediments or highly conducting layers, E and G.

The Rôle of Water in Relation to Cancer

From an electrical point of view, pure water, *i.e.*, H_2O containing nothing but hydrogen and oxygen, is a very good insulator, and the same applies to soft water found in sandy soils. On the other hand, waters containing salts, such as sea-water and mineral waters, act more or less as conductors, and at times they may prove to be very good conductors. It is this ' polymorphism ' of water which may account for the fact that certain waters seem to be associated with the incidence of cancer while others have apparently no influence at all.

Many districts and cities built on the banks of rivers do not necessarily have a high incidence of cancer. In Paris, near the Seine, both a high and a low incidence of cancer have been observed, which seems to prove the absence of correlation. A city, such as Antwerp, with a low incidence of cancer, is built on the banks of a great river, near a vast estuary, while Geneva, which also has a low cancer incidence, is built in close proximity to a large lake on an alluvial bed. But, on the other hand, cities like Nancy, Saint-Etienne and Strassburg, which are built on the banks of small rivers, have a high incidence of cancer.

These observations seem to show that water does not play a part in the incidence of cancer except when its electric constants and the form of its volume (water-beds, etc.) are of such a nature as to affect the field of cosmic radiation which may break up the equilibrium of cellular oscillation.

In the light of these facts we are in a position to realise why many reputable writers have often drawn attention to the existence of "cancer houses," "cancer streets," "cancer villages," and "cancer districts."[1] We have indicated the preponderant part played by the nature of the soil in the localisation of cancer. It may easily be shown that the soil of such localities contains at a variable depth certain layers acting as particularly good conductors: plastic clay, ferruginous and arsenical ores, carboniferous and other layers.

An eminent cancer research worker, Dr. Hartmann, has drawn attention to the fact that a medical observer had been impressed by the high incidence of cancer in the Ognon Valley.[2] Now this river flows in a bed of Jurassic formation where conducting plastic clay abounds.

In the matter of the specific influence of water on the incidence of cancer, I have suggested the following explanation based on the laws of electricity.

Water, which is neutral in a state of purity, takes on the conducting properties of substances with which it comes into contact, even as, from a chemical point of view, aqueous solutions show the properties, either acid or alkaline, of dissolved substances.

Again, mineral waters contain in solution mineral substances such as sulphur, carbonates and bicarbonates, iron and arsenical salts, etc., which are derived from various geological formations. Such waters possess, at their exit from the soil, the same chemical and electrical constants as the soil from which they emerge. If further evidence is required it may suffice to point out that oscillations characteristic of mineral waters are shown by the radio-activity of these waters in the immediate neighbourhood of the springs. Radio-activity results from the oscillatory disequilibrium of certain mineral substances

[1] Orthodox medical writers usually dismiss the question of "cancer houses" as being unworthy of serious attention but recent statistics in Budapest have shown that there actually are so-called "cancer streets" and "cancer houses" where the number of cancer cases is strikingly large. [Foreign Letters—*Journal of American Medical Association*, July 13th, 1935 (Translator).]

[2] H. Hartmann, "Rapport sur la contribution à l'étude étiologique du cancer par M. Chaton," *Bulletin de l'Académie de Médecine*, vol. 11, Mars 5th, 1927, p. 348.

which, at their exit from the soil, do not oscillate in harmony with cosmic waves. From the earliest times physicians have observed that the efficacy of mineral waters was particularly noticeable near the springs owing to the harmony which is then attained between the cellular oscillation of the individual, the radio-activity of the spring and cosmic radiation.

Moreover, my views on this point have been confirmed by many observations.[1] Hoffman observed that while the cancer mortality reached an average of 0·85 per 1,000 and even 1·199 at Boston in 1915, the corresponding figure for Memphis was only 0·467.

In his monograph on " Cancer and Water " [2] Dr. Shannon has shown that the city of Memphis is supplied with water from artesian wells situated in the soil of the city itself. Dr. Shannon attributes the low incidence of cancer in Memphis to the water of these artesian wells which, according to him, is free from protozoal organisms. But no one has yet succeeded in proving that cancer is caused by the presence of protozoa in water.

In the light of our theory, the water of these artesian wells is a mineral water possessing the same characteristics as the soil on which the inhabitants of Memphis live. As they use this water for both external and internal purposes, these people are thus automatically placed under such conditions that their cells have the same electrical and chemical constants as the soil of their habitat, and consequently they may be said to be " in resonance " with the local field of cosmic radiation.

In France, similar observations appear to corroborate these facts. At Luxeuil, Dr. Thomas observed an almost total absence of cancer. It seems that, owing to the scarcity of potable water the inhabitants of this locality drink only the mineral water of the Spa establishment, obtained from the depths of the local soil.

Recently the same observation concerning the relative absence of cancer was made at Châtel-Guyon. A Com-

[1] F. L. Hoffman, " The Mortality from Cancer throughout the World." Newark, N.J., 1915.

[2] J. W. Shannon, " Cancer and Water : a Study of the Nature, Causation and Prevention of Cancer." San Diego, California, 1917.

mission of French and foreign cancerologists paid a visit to this famous resort known for its water, in order to investigate the causes of the reported low incidence of cancer. Now it is known that the water supply of this town does not come from a distant source, but is derived from a local site, Mont Chaluset. The explanation suggested for the phenomenon observed at Memphis and Luxeuil is also valid for Châtel-Guyon. Furthermore, we may remark that the water supply of Geneva is drawn from the depths of the Lake of Geneva and therefore possesses the same electrical constants as the lake and the soil themselves. The cancer density in Geneva is said to be distinctly low (0·50 per 1,000) which would seem to confirm our original explanation.

In this connexion another significant observation was made by Dr. Simeray,[1] who reported that the population of an entire village was free from cancer as long as they made exclusive use of water drawn from wells sunk for this purpose. But when the local authorities decided to obtain their water supply from a source outside the locality and to give up the use of the wells, a series of cases of cancer occurred in the village. In this case the appearance of cancer seemed to coincide with the utilisation of a distant water supply which did not possess the same electrical constants as the soil of the locality and consequently caused in the villagers' bodily cells a state of oscillatory disequilibrium in relation to cosmic radiation.

I was able to verify Dr. Simeray's observation personally in the case of two neighbouring localities, Thiais and Orly (Seine-et-Oise). Both are situated on the same kind of soil —fresh-water limestone of Brie—which is a fairly good conductor and therefore characteristic of a high cancer density. But the density of cancer for Thiais is 3·36 per 1,000 and only 0·36 for Orly. As this case did not appear to be in accord with my theory on the subject, I decided to investigate the conditions myself with the assistance of the local authorities. I found that at Thiais the water supply came from the Seine, drawn at Alfortville, whereas at Orly, the inhabitants drew their water from their own wells situated in the centre of the locality.

[1] Session of the Académie de Médecine, March 15th, 1927.

CHAPTER X

THERAPEUTICS OF CELLULAR OSCILLATION

In the preceding chapters I have shown that a living organism, whether animal or plant, is comparable to a system of high frequency oscillating circuits consisting of cells which are themselves elementary oscillators.

I have indicated the nature of radiation in living beings and how different rays influence them. I have particularly stressed the rôle of cosmic radiation and how it is influenced by various physical factors such as the conductivity of the soil and the effect of astral radiation, resulting in interference phenomena.

All the investigations I have undertaken appear to confirm the fact that diseases are the outcome of oscillatory disequilibrium resulting from (1) certain modifications in the field of cosmic waves in consequence of interference through a secondary field at the surface of the soil, (2) from an astral radiation (solar, lunar) or else, which comes to the same thing, (3) from modifications of the electrical constants of the living cell.

Thus I have been led to evolve a new therapy whose object was to re-establish the cellular oscillatory equilibrium that had been disturbed by disease. According to the nature of the case, it may be advisable to act directly on the diseased organism by means of *biomagnomobile* substances or substances capable of restoring to the cell its appropriate electromagnetic constants (capacity, self-inductance and resistance of nuclear oscillating circuit) ; or it may be more expedient to act indirectly by modifying the field of cosmic waves around the patient by means of some suitable radio-electrical apparatus.

The object of this method is to regulate the electromagnetic field within organic tissues, chiefly by reconstituting the positive and negative fragments of every cellular nucleus, a process involving the utilisation of *biomagnomobile* substances. And lastly, we know that the magnetic field is due to a rotatory motion of electrons which is a particular kind of oscillation.

My researches on cancer have led me to the conclusion that this terrible disease is least prevalent in localities where living organisms are in harmony, that is to say in oscillatory equilibrium with the soil of their habitat, as I have indicated before.

It seems that we have here a universal principle which may prove useful in therapeutics, and it is even more a principle of general hygiene than a therapeutic principle.

In my monograph on the "Contribution to the Etiology of Cancer," I showed that certain favourable conditions were established when the inhabitants made use of water drawn from the depths of the soil upon which they lived. I am convinced that if people could subsist exclusively on fruits and vegetables grown in gardens attached to their houses, and made use of water drawn from wells sunk close by, cancer and most other diseases would become far less prevalent. Do we not often hear of country people reaching an advanced age in spite of deplorable hygienic conditions under which they live? This longevity may be accounted for by the fact that these country people are compelled to make use of their local water supply and to live on their own produce.

The disadvantages of modern water supplies might be overcome in cities by sinking artesian wells, such as those existing in Paris in the Place Lamartine, the Avenue de Breteuil and the Bois de Boulogne. As for the new artesian well of the Rue Blomet, it would be infinitely better if this water were used for household purposes rather than for a swimming pool.

When local living conditions are exceptionally bad or variable, it is possible, as I have shown, to re-establish, or rather "tune up" the electric constants of the cell by means of appropriate substances in harmony with the physical and chemical nature of the soil of the habitat. These substances could be administered by hypodermic injection or, preferably, by the oral route. At night time the sleeper might be connected with the soil by means of an appropriate earth connection, and in the daytime, footwear might be used to hold a metallic plate in the sole or heel, thus establishing electrical contact between the foot and the soil. In the majority of cases it seems more rational and more efficacious to resort to electrical methods such as filtration of the field of cosmic waves in the immediate vicinity of the individual.

I have also recommended the use of special radio-electrical contrivances such as metallic antennæ fixed up in flats or outside houses, earth-connections, metallic grids and, preferably, appropriate oscillating circuits.

The filtration of cosmic radiations systematically carried out by these oscillating circuits is, in point of fact, accomplished naturally by radiations of longer wavelength, such as luminous rays, ultra-violet rays, X-rays and radium emanations. This accounts for cases successfully treated by heliotherapy, actinotherapy, radiotherapy and radio-active substances.

My experimental work has confirmed the soundness of the foregoing principles. Let us recall to mind the experiments with geraniums inoculated with *Bacterium tumefaciens* and treated by the radiations of my Radio-cellulo-oscillator, with the result that the plants were cured after a few applications. Since then I have shown that disease occurs owing to oscillatory disequilibrium brought about by excess of cosmic waves. The ultra-short waves emitted by the Radio-cellulo-oscillator reconstitute, by interference, the field of cosmic radiation which thus acquires an appropriate value, the same result being achieved by the intervention of luminous rays, ultra-violet and radio-active rays.

In a preceding chapter I indicated, in regard to the nature of radiant energy, how I obtained the same curative result with geraniums by eliminating the Radio-cellulo-oscillator and substituting for it the simple device of a copper spiral encircling the plants. This spiral is the simplest and most general form of an oscillating circuit which I advocate for the filtration of cosmic waves in connexion with the treatment of various diseases, including cancer.

The results I obtained in treating these plants by means of an oscillating circuit were far beyond my expectations.

Professor d'Arsonval, who presented my communication to the Académie des Sciences, drew attention to the fact that at the beginning of January, 1925, I had set up an oscillating circuit consisting of a copper spiral suspended in the air and kept in position by means of an ebonite support introduced into one of the flower pots containing the geraniums inoculated with cancer on December 4th, 1924. On January 30th, 1925, the tumour was developing normally, but the plant continued to grow without showing signs of decay, whereas all the control plants had perished as a result of the tumour they bore. At the end of February, 1925, the treated plant was cured and the necrosed tumour had fallen off. On March 23rd, 1928, the same plant, still encircled by its oscillating circuit, was photographed (Plate VII). Comparison of the photographs of January 30th, 1925 (page 103), and of March 23rd, 1928 (page 144), reduced to the same scale, gives an idea of the extraordinary development of the plant which, in three years, reached a height of 1·40 metres, *i.e.*, about 4½ feet. This geranium is still flourishing, even in winter, and appears to be in excellent condition. It should be borne in mind that tumours due to *Bacterium tumefaciens* usually cause cachexia and death, even after surgical removal.

Since this first experiment, many investigations in the same field, in conformity with my methods, have been carried out in France, Italy and America. I, myself, have extended my researches on plants to animals and human beings, and it has given me great encouragement to know that my methods have been successfully applied by eminent workers in laboratories and clinics. Among the numerous reports published concerning these experiments, special mention must be made of the report presented at the Congress of Radiology in Florence (May, 1928) by the eminent authority on cancer, Professor Sordello Attilj, of the Hospital San Spirito in Sassia, Rome. Only a brief summary of this report can be given here.[1] Professor Attilj made extensive use of my open oscillating circuits

[1] A complete account of this report was published in my article on "The Theory of Cancer based on the Geological Nature of the Soil." (*Revue générale des Sciences*, Octobre 15th, 1928.)

PLATE VII

Photograph of Geranium three years after treatment with oscillating circuit showing remarkable development of plant. Two untreated control plants are shown beside it. This is the same geranium as in Plate IV, on page 103. (*Surgical Clinic of Salpêtrière, Paris.*)

which I recommended in the form of collars, bracelets, belts, etc.

The most important observations of Professor Attilj, appearing in the report in question, concern six patients—five of whom were suffering from cancer and the sixth from polysarcia (excessive corpulence). All these cases of cancer exhibit marked individual differences.

1. Patient, aged 78, suffering from epithelioma (ulcerated) of the floor of the mouth, with sub-maxillary metastases.

2. Patient, aged 25, suffering from recurrent sarcoma of left hand.

3. Patient, aged 28, suffering from recurrent sarcoma of right breast.

4. Patient, aged 60, suffering from epithelioma (ulcerated) of genital organs.

5. Patient, aged 40, suffering from severe post-operative pains with small metastases in scar on breast.

It will be noticed at the outset that three cases of cancer are complicated by recurrence or secondary manifestations (metastases) which constitute aggravating conditions. Nevertheless, a few weeks after application of oscillating circuits, Professor Attilj noted a diminution of pain, a progressive resorption of lesions and disappearance of induration of tumours. In the majority of cases the painful formication (" pins and needles ") accompanying the development of tumours ceased when the oscillating circuit was applied. The sixth case concerning the patient suffering from polysarcia is perhaps the most interesting of the series. Weighing 120 kilograms, the patient was suffering from lancinating pains in the lumbar region and moved with such difficulty that it took her three to four minutes to rise from the sitting position. Three days after the application of the oscillating circuit (a belt in this case) the pains disappeared, the patient regained her appetite, so much so that at the end of three months' treatment she was able to move with ease and resume her normal activities.

Professor Attilj summed up as follows :

" The small number of cases treated which represents only the beginning of a method of treatment awaiting further development, shows that the use of Lakhovsky's oscillating circuit is really effective. When we bear in mind the tragic fate of cancer patients doomed to die, often in great pain, while at the same time their organs are gravely affected by the disease, it must be admitted that anything that can relieve such distressing symptoms is a great blessing to the suffering patients."

Professor Attilj admits the efficacy of open oscillating circuits for re-establishing cellular oscillating equilibrium, not only in cancerous patients, but also in patients suffering from cardio-vascular and nutritional affections.

For some time past I have made similar observations myself and have collected a great number of reports from practitioners who have cast aside their preconceived ideas in the interest of science, and have experimented with my methods of treatment. [For particulars and photographs of cases treated, see Appendix.]

Generally speaking, the following conditions have been most often dealt with by practitioners :

Insomnia, due to overwork or following an illness, is successfully treated.

Pain associated with various affections is generally reduced, sometimes eliminated, even in cases of cancer.

Patients have noticed a sensation of warmth due to activation of the circulation. Blood analysis shows an increase of red corpuscles. Anæmia and cold extremities are thus amenable to our methods of treatment.

The gastro-intestinal functions are stimulated and gastric acidity is reduced while intestinal atony and pains accompanying digestion also show a favourable response.

In deaf patients, an improvement has been observed.

Other signs of improvement include better appetite, increase in weight and an appearance of rejuvenation, often distinctly marked.

Lastly, attention must be drawn to the interesting observations made by a distinguished French professor who experimented with my methods in one of the great Paris hospitals. The patients under treatment were subjected to strict examination. Once a week the weight was noted, the blood analysed and blood pressure recorded. While these experiments were in progress the professor noticed that during a period of about eight days the improvement previously observed came to a definite standstill in all the patients. He deduced from this general phenomenon that an external cause was operating. In looking at the calendar he observed that this abnormal period coincided with the full phase of the moon.

From the point of view of my theory this phenomenon may be explained as follows. We know that the moon, in common with all sources of radiation, has the power of causing considerable variations in the field of cosmic waves, a subject fully dealt with in my work "L'Universion." Moreover, the effect of the oscillating circuit is to absorb any excess of cosmic waves which are responsible for the oscillatory disequilibrium of the cells. As the moon modifies the field of these waves, this interference has repercussions on the absorption of the oscillating circuit whose action is diminished. Thus we observe that the effect of an oscillating circuit worn by a patient is in close relationship with the field of cosmic waves. In cases in which this effect is inadequate, the desired result may be obtained by making use of several circuits (collars, bracelets, belts).

As a general rule, I have observed that in all the patients wearing oscillating circuits and living on highly conducting soils, that is to say naturally carcinogenic (cancer-producing) such as Grenelle, Javel, Auteuil, Neuilly, the action of the circuit is immediate and rapid, whereas in patients living on insulating soils, such as Dauphiné, the Champs-Elysées, Gaillon, Monceau, this action is much slower, and its effects are not manifested until a certain time has elapsed.

Thus the action of an oscillating circuit, being closely connected with the intensity of the field of cosmic waves, gives rise to the paradoxical conclusion that thanks to the use of this circuit, the worst soils, from the point of view of health, finally turn out to be the best. The oscillating circuit (collar, belt, etc.) acts by regulating the incidence of cosmic waves, thus re-establishing, automatically and naturally, the oscillatory equilibrium.

We are justified in concluding, therefore, that the application of open oscillating circuits succeeds in arresting the development of cancer, even in the most advanced stages, while pain is eliminated and the dreaded disease sometimes conquered.

Finally, similar gratifying results have been obtained in the treatment of many other diseases which apparently have no connexion with cancer. Thus it may be claimed, *a fortiori*, that oscillating circuits, in absorbing excess of cosmic waves, may prove to be a means of preventing disease worthy of consideration.

I am hopeful that, in future, all diseases afflicting mankind may be prevented and successfully treated.

CHAPTER XI

ORIGIN OF LIFE

Condensation of Water Vapour and Mineral Elements—Influence of Cosmic Radiations on Orientation of Cellular Elements—Constitution of Electric Oscillating Circuit of the Cell—Characteristic Elements of Living Species—Problem of Heredity—Infinitesimal Value of Radiant Energy—Induction in Fixed and Oscillating Fields —Induction in Electromagnetic Fields within the Cell.

Condensation of Water Vapour and Mineral Elements. In the geological epochs, when life had not yet appeared on the surface of the earth, our world which had stored, at a certain time, the condensation of all the water vapour in the atmosphere, was partially or totally covered with oceans.

The elements and various chemical compounds, dissociated under the action of heat, then subsequently condensed, were found scattered everywhere. They are still found, almost without exception, in sea-water, whose analysis reveals a great complexity: chlorides, bromides, iodides, sulphates and most salts of the principal metals: sodium, potassium, magnesium and many others. It is entirely owing to humidity in the neighbourhood of the sea or in the sea itself, that life emerged and that the first protozoon appeared.

As biological science has established the fact that the first phase of life is the cell, I propose showing how the primordial cell was formed by referring to my theory of cellular oscillation.

It is important to bear in mind that salts, simple bodies and other chemical compounds which existed in a state of great dilution in the midst of vast masses of water and saturated vapours, were, in consequence, strongly dissociated and ionised, in the form of atoms and molecules, more or less electrified. Thus every droplet of water formed a tiny microcosm containing, in a state of extreme dilution, a great variety of chemical elements. Hence it must never be lost sight of that humidity is essential to life ; it was the first condition for the appearance of life on earth.

Influence of Cosmic Radiations on Orientation of Cellular Elements. The causes determining the generation of cosmic waves were already existing when the earth appeared in the universe. The radiations which generate cosmic waves, whether from the sun or from other stars, have remained unaltered. But our earth, at that time, as even now, must have been charged with negative electricity.

The process of the appearance of life may conceivably have been as follows : under the action of electromagnetic radiations of cosmic origin, certain molecules of chemical compounds and certain atoms of simple elements, contained in globules of water, were orientated along lines of force of an electric field generated by some astral body, charged " positively " while the earth was charged " negatively."

Let us note that owing to the multiplicity of astral electric fields, orientation of molecules could have been effected just as well along lines of force coming from the sun as from the moon, Mars, Jupiter, or any other planet or astral body.

Again, these molecules of conducting substances,. containing iron, potassium, iodine, chlorine, and various chemical combinations, were automatically grouped under the influence of chemical affinity or electrostatic forces. They began to form along a certain line of force a small agglomeration of electrified molecules to which other molecules were attracted. These unions, however, occurred according to a determined direction, that of the line of electromagnetic force which, arising out of celestial space, reached the earth, negatively charged, as modern science has shown.

These groups of conducting molecules were thus orientated and joined together in the form of an extremely short curved rod.

Around this " bait," a certain number of molecules from insulating substances came to be fixed, possibly owing to the force of gravity, and formed, as it were, a sheath enveloping the original agglomeration of conducting molecules.

Constitution of Electric Oscillating Circuit of the Cell. Owing to the rotation of the earth, the orientation of agglomerated molecules was subjected to deviation and as a result of its rotatory motion the earth thus played a part, at the end of twenty-four hours, or even after a few days,

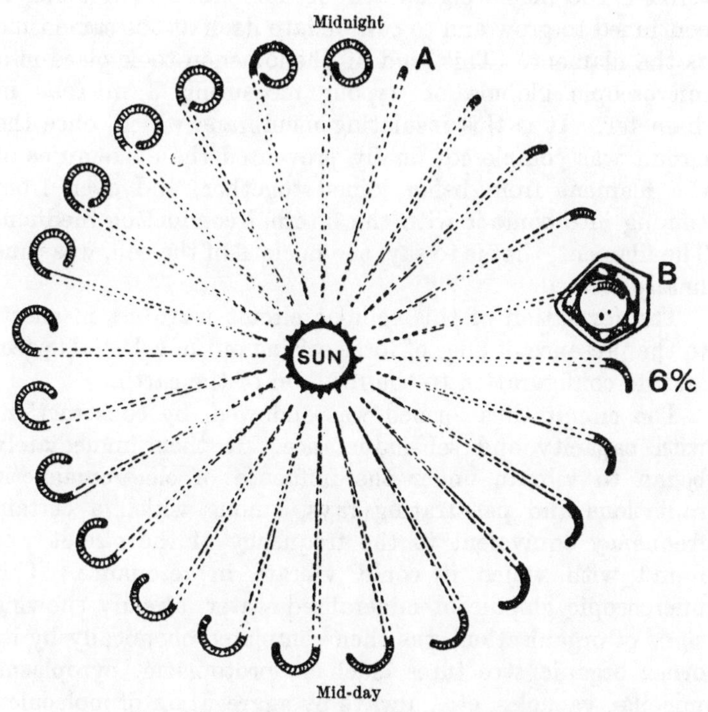

FIG. 22. FORMATION OF PRIMORDIAL CELL.

Starting from A at the top of the figure and going clockwise round the sun, it will be observed how the filament, developing along a line of force emanating from the sun, has acquired its incurved shape in consequence of the rotation of the earth. The filament thus had its circuit completed in twenty-four hours. The cell at B shows a phase of its formation.

This diagram is purely schematic. Actually the cellular nuclei are much more irregular than they are depicted here. The reason for this is clear. In the course of the rotation of the earth on its own axis in twenty-four hours, astronomical conditions are bound to vary and the line of force emanating from the sun will clash with other lines of force originating from various stellar bodies. This results in a momentary stop or a deviation in the original line of force during the formation of the filament-nucleus whose disposition is thus rendered irregular.

in the formation of a filament no longer rectilinear but curved, and at times, having the shape of a tangled cluster (Fig. 22). The new parts of this filament were consequently formed along the line of magnetic force, invariable in direction, while the parts already formed were swept along by the motion of the earth. As this conducting filament was being formed, the insulating sheath or membrane enveloping it continued to grow and to consolidate itself at the same time as the filament. This kind of phenomenon took place in a microscopic globule of vapour measuring 3 microns in diameter. It is this insulating membrane which, once the circuit was completed, finally prevented the extremities of the filament from being joined together, and even from coming into contact with the internal conducting medium. The filament, that is to say, the nucleus of the cell, was thus finally formed.

The formation of this cellular circuit was due, in short, to the presence of lines of force arising out of celestial space, and its configuration to the rotation of the earth.

The circuit thus formed was endowed, by construction, with capacity and self-inductance. It then immediately began to vibrate under the influence of electromagnetic radiations and penetrating rays, among which a certain frequency equivalent to the frequency of the circuit was found with which it could vibrate in resonance. This microscopic globule of mineralised water, already showing signs of organisation, was then completed chemically by its other organic structures, such as protoplasm, cytoplasm, micellæ, vacuoles, etc., always by aggregation of molecules. And as it was vibrating and radiating, this globule was living and the cell was born.

Characteristic Elements of Living Species. As a result of this formation every cell, or at least, every species characterised by cells possessing nucleus and protoplasm, began to oscillate with a frequency and wavelength determined by the dimensions of its filament. Thus by virtue of the form and dimensions of the filament, every cell, like every microbe, possesses its own wavelength which is characteristic of its species. But all these cellular wavelengths, although widely different, are of the same order of magnitude and approximate to one another within a narrow zone of the whole range of vibrations.

According to my theory, this definition of cellular species

involves one of the following consequences. If, by any process, we succeed in modifying the duration of the formation of a cell, which implies modifying the constitution of its filament or of its conductive capacity, either by means of chemical elements or by electromagnetic methods, we modify at the same time its frequency of vibration, and consequently, the species of the cell as well as its particular characteristics.

This sequence of events probably occurs in cases of cancer, diseases of old age, etc. Transmutation of cells would thus be accomplished.

Experimental evidence gives support to this view. Furthermore, a similar state of affairs may occur in the action of certain medicaments of mineral, vegetable or animal origin, which are intended to cure certain conditions by reinforcing the conductivity of the nucleus, or modifying its chemical nature, the nucleus being of primary importance in the process of oscillatory disequilibrium.

Cell Differentiation and Heredity. The most diverse hypotheses have been enunciated on the constitution of protoplasm.

According to Naegeli, matter is composed of units to which he gave the name of micellæ. Other scientists, such as Darwin, Haeckel, Spencer, Hertwig, de Vries, Wiesner, have been led to postulate the existence of a physiological unit of a higher grade than micellæ, *i.e.*, the *idioblast*. The sum of idioblasts constitute the *idioplasm*. Hertwig states that the hereditary substance is not located in the protoplasm but rather in the nucleus and has adopted Pflüger's concept of the *isotropy of the ovum*, that is to say, that the ovum is homogeneous and none of its parts corresponds beforehand to any part of the future animal. Weismann propounded the theory of "ancestral plasma."[1] The problem of the specific differentiation of cellular elements has given rise to a great number of hypothetical solutions including the theories of His, Hansemann, Hertwig, Naegeli, de Vries, etc.

According to my views, the hereditary substance is

[1] Weismann was the author of the germ-plasm theory of heredity which denied the transmission of acquired characteristics. (Translator.)

located neither in the protoplasm, idioplasm nor in the micellæ, but actually in the nucleus ; and the specific differentiation of the nucleus is due to its power of vibrating in accordance with a wavelength determined by the diameter of the circuits which constitute it and the value of the nuclear electrical capacity. In procreation, the male or female cell which proves to be dominant is that whose wavelength approximates most closely to the normal standard typical of its sex. This may account for hereditary phenomena brought about by nuclei whose diameter does not vary for generations, their wavelengths and the chemical composition of protoplasm that form capacity remaining consequently unchanged. This may also account for the recurrence of qualities, defects, resemblances, etc., through many generations, in short, what is known as atavism.

Infinitesimal Value of Oscillating Cellular Energy. In the course of this work I have already raised the question : " Whence comes the energy of cellular radiation ? " It is this question which I propose answering now by way of concluding the formal statement of my theory.

Owing to the microscopic dimensions of cells and their filaments, dimensions measuring only fractions of microns, it follows that the oscillation of such a circuit requires but an extremely small amount of energy. It is difficult to imagine the infinitesimal quantity of this energy, but the imponderable amount of force brought into play in the course of these oscillations does not preclude the far-reaching effects of these ultra-short waves, owing to the considerable amount of induction attainable with such high frequencies. Let us call to mind, for instance, the vast range available to wireless stations making use of so-called short waves which are actually long waves in comparison with cellular oscillations. For such broadcasting a power of a few dozen watts is sufficient, and it has even been possible to reduce this to 1 watt or less while operating in a radius of more than 2,000 kilometres.

Some physicists have experimented with high frequency waves of the order of a hundredth, and even a thousandth, watt. In the experiments of Nichols and Tear, for the generation of electromagnetic waves of 300 microns, the energy of these radiations was so attenuated that their measurement necessitated a special optical method.

Thus a certain imaginative effort is required to appreciate the greatly attenuated energy that makes the circuits of our cells oscillate, whose structure is perceptible only under the microscope at a magnification of 300 to 500.

We shall not attempt to calculate this energy ; suffice it to say that it is infinitely small for each circuit. We have seen that the wavelength of cosmic waves is extremely small and that atmospheric radiant energy is sufficient to cause cells to oscillate. When a Hertzian wave emitted in Australia, for instance, with a power of a few dozen watts, is transmitted in all directions and ultimately received in Europe by a small aerial, the high frequency energy picked up by the receiving aerial is infinitesimal. It is all the more so as the energy decreases theoretically in inverse ratio to the square of the distance, and practically with far greater rapidity.

Induction in Fixed Oscillating Fields. How is it possible that such a receiving aerial, picking up so little energy, can yet oscillate in its turn sufficiently to activate another far distant aerial ? This is largely due to the very high frequency of these short waves, whose attenuated length approximates more closely to the wavelength of cosmic radiations than to that of long waves.

We know that the process of wave reception is as follows : the receiving aerial is situated in a variable electromagnetic field created by the waves which are propagated from the transmitter. It is this variable high frequency electromagnetic field which, by induction, generates oscillating electric currents of the same frequency in that aerial. It is also owing to this same mechanism that our cells oscillate, and I shall show whence the necessary energy is derived.

At this stage it may be useful to draw attention to two essential conditions relative to induction phenomena bearing on sustained oscillations.

In order to bring about generation of oscillating electric currents in a circuit, it is necessary that the following conditions should be fulfilled.

1. Existence of an electric circuit capable of oscillating (circuit possessing self-inductance and capacity).

2. Existence of an external cause capable of making the circuit oscillate.

We have seen that the first condition was fulfilled in every cell. With regard to the second condition, the phenomenon of oscillation may be due to a great variety of causes. In any case it is sufficient that the self-inductance of the circuit in question should be influenced by an oscillating magnetic field or that the capacity should be situated in an oscillating electric field.

Each of these two induction phenomena, electric or magnetic, may itself be brought about in two ways.

In the first case, the self-inductance of the circuit is *fixed* and the external magnetic field (or the electric field in the case of a condenser) is variable. This variation of the field then produces, by induction in the circuit, currents whose frequency corresponds exactly with its own wavelength. The effect may actually be determined by a multitude of fields, each having its own frequency, induction being produced only by the field whose wavelength coincides with that of the circuit.

In the second case, self-inductance is *mobile*, and is displaced with very great speed in the magnetic field. The action of an electric field on the capacity of the circuit would take place on similar lines.

The electric or magnetic field in question may be variable in regard to time and show exactly the same frequency as the induced currents in the circuit. Or else this field may be variable in regard to space, for instance, an undulating field having a fixed value on which discontinuities or interruptions are superimposed. Or yet the field may be fixed though the oscillating circuit itself is mobile. It is upon these phenomena that the construction of industrial alternators is based; in certain cases the revolving part is constituted by the continuous current induction circuit whose magnetic poles acquire a high velocity in the presence of the fixed coils of the induced circuit. The induced alternating currents then arise from the circuits of the fixed part which, by the rotation of the revolving inductor, are subjected to variable magnetic fields.

The same principles apply to a frame aerial; the spirals of the frame act by induction, like the secondary circuit of a transformer, whose primary circuit corresponds to a transmitting aerial. Induction is produced by the variable magnetic field propagated by the waves issued from the transmitter. It is owing to the same process that the radiant energy of cosmic waves activates our own cells.

Induction of Electromagnetic Fields within the Cell. We have seen that living cells possess oscillating circuits constituted by filaments. Now all these cells are set in motion in space, impelled by the motion of the earth, at a velocity of 27 kilometres per minute at the Equator. The question is in what particular field do these cells revolve? Evidently not in the terrestrial electromagnetic fields, since these fields are swept along at the same time as the cells by the same rotatory motion. The cells revolve in variable electromagnetic fields generated by a source external to the earth, that is to say within the field of atmospheric radiations comprising a complete range of frequencies as typified by cosmic radiations emanating from the sun, the Milky Way and the immensity of celestial space.

Finally, the existence of variable electric and magnetic fields of multiple frequencies arising out of space shows that all the energy of radiation at the earth's surface comes, in the last analysis, from electromagnetic induction brought about by the rotation of the earth in space.

Let us now consider the relations existing between the chemical composition of the cell and its radiation. We know that living beings, animals and plants, in a word every living cell, contain all the chemical atoms in their great complexity. As I stated before on the subject of cell differentiation and heredity, various names have been given to the elementary units of the cell and protoplasm, such as micellæ, idioplasm, mitochondria, etc. From my point of view I prefer to describe them as *biomagnomobile* units so as to stress their biological origin, their essential mobility and the electromagnetic element which charges them with vital energy.

Let us take, for example, the process of electro-plating in which two metallic electrodes are immersed in a conducting liquid. The metallic atoms are swept along by the current; they leave one electrode to be deposited on the other electrode, this being due to elementary electrostatic charges, each atom being directed by electrons moving from one pole to the other. When the current fails the motion of atoms ceases.

In the case of living cells the number of these particles constituting a single cell is incalculable. Thus, according to Raphael Dubois, it would take 250 million years, provided it were possible to count at the rate of one million per second, to estimate the sum total of units contained in the egg of a silk-worm. Whatever the number may be, these units are incessantly moving in our organism; thus a cell in the brain may ask a cell in the stomach, for example, to supply it with a few hundred trillions of those biomagnomobile units (derived from phosphorus, chlorine, iron, etc.) which circulate in all parts of our bodies. Molecules are first brought into the blood by foodstuffs or formed within the organism from simple elements. And all these molecules are set in motion, attracted or repulsed, by the play of cellular oscillations, as in the motion of charged electrical particles in the process of electro-plating. Moreover, the organism consists only of *living biomagnomobile* units, in a state of incessant chemical and electromagnetic activity. All functional activities can only be carried out as a result of the harmony and general organisation of the cells and of their oscillations originating from cellular nuclei. It is this general harmony which determines the particular position of every molecule. With regard to the necessary energy, this comes from the electrical vibration of the cells, energised in their turn by cosmic waves.

At this stage the question may be asked: "What about toxins?"

Toxins are the waste products of cells and of dead microbes. As they are no longer living, and thus constitute inert matter, these waste products neutralise the oscillatory movement of neighbouring cells and weaken them or cause their destruction. These inert particles attract living particles ; in any case, their proximity modifies the electrical capacity of living cells which can no longer oscillate in accordance with their specific frequency ; hence disease or death.

In this connexion let us consider the action of the microbe on the cell from a biological point of view. First let us point out that the microbe does not attack living cells directly, but only indirectly by induction, as we shall show later. Chemical analysis of microbes and cells shows them to be remarkably similar in composition. It seems, therefore, *a priori*, that it is difficult, from a chemical point of view, to account for the action of the microbe. But if we investigate the chemical composition of microbes and cells respectively, the distribution of the different substances enables us to solve the problem of this " war of radiations " which I mentioned in an earlier chapter.

We know that the constituents of living cells and microbes may be classified into three categories : nitrogenous substances, ternary substances,[1] and mineral substances. Thus, for instance, analysis of a cell of the fruit-bearing part of *Æthalium septicum* shows the following proportions :

Nitrogenous substances	30
Ternary substances . . .	41
Mineral substances . . .	29
Total . . .	100

According to Henneguy, in nitrogenous substances are found : plastin, vitellin, myosin, peptones, pepsins, lecithin, guanine, xanthin and ammonium carbonate. In ternary substances : paracholesterol, a special resin, a yellow pigment, amylodextrin, a non-reducing sugar, fatty acids and neutral fats. In mineral substances : lime combined with fatty acids and other organic acids, such as lactic acid, acetic acid, formic acid, oxalic acid, phosphoric acid, carbonic acid, sulphuric acid ; phosphates of potassium and magnesium, chloride of sodium and iron salts.

[1] Ternary is a term indicating that chemical compounds are made up of three elements or radicals. (Translator.)

Generally speaking, all chemical substances present in sea-water are found in the human organism.

From the point of view of my theory of cellular oscillation, all the substances enumerated above may be divided into two categories :

1. Conducting substances.
2. Insulating substances.

As a general rule, insulating substances are found in nitrogenous and ternary compounds and conducting substances in compounds containing mineral salts. Thus, for instance, plastin, paracholesterol, resin and certain fats are insulating, while most minerals, and particularly the salts (sulphates, phosphates, chlorides, of sodium, magnesium, iron, etc.) are more or less conducting.

In the light of this classification we shall see how the microbe may, by induction, modify cellular oscillation. Let us recall that oscillation in a circuit depends on its conductivity (electrical resistance) and on its permeability to waves (specific inductive power and capacity). Returning to the cell of *Æthalium septicum*, we have seen that its chemical composition was as follows : nitrogenous substances 30 ; ternary substances 41 (most of these being insulating) ; mineral substances 29 (most of these being conducting).

Let us suppose that this cell is attacked by a microbe whose mineral ratio is 40 instead of 29. Its oscillating power and consequently its frequency, are not the same as those of the cell. Thus, by induction, the microbe modifies the oscillation of the cell, which results in its destruction and death. Again, the cell, instead of dividing normally by karyokinesis into daughter cells, divides according to the frequency of the microbe, that is to say into cells typical of the microbe. In the absence of a microbe, if the nucleus of the cell is too powerful a conductor (excess of iron and phosphorus derived from globulins), and if the external agent (excess of cosmic waves) causes a too rapid division of cells, we may find that the healthy cell will be transformed into a neoplastic cell (cancer).

The foregoing facts show that in a healthy organism every tissue must contain, in constant proportions, conducting and insulating constituents which I have named *biomagnomobile* units.

The question now arises how the distribution of these units in the organism is effected so as to bring to the membrane of the nucleus the insulating substances, and to the filament the conducting substances.

It is essentially due to the energy of its own oscillation that the cell is able to summon for its needs all these insulating and conducting substances which are distributed to the locations where they are required for the maintenance of the life of the cell itself. Similarly, in the electro-plating process the substances and the strength of the current are adjusted so as to obtain the desired effect, according to the nature of the metal employed.

Such is the final remark with which I conclude the formulation of my theory.

My experiments in the field of radiobiology are now established facts which cannot be accounted for by the classical theories of science, whereas my new theory provides the necessary explanation.

In conclusion, my theory may be summarised in the form of this threefold principle :

Life is created by radiation,
Maintained by radiation,
Destroyed by oscillatory disequilibrium.

Be that as it may, I believe I have opened up a new field of research which should prove particularly fruitful to biologists. No one can predict what the future has in store for us in this field ; in any case, I hope that the ultimate result will benefit suffering humanity.

CONCLUSION

In concluding the presentation of my theory and of its practical applications, I wish to appeal to physicists, research workers, to all men of science in general, for in them lies the source of all progress. It is they, in particular, who have achieved the modern marvel of wireless. If anyone had predicted, forty years ago, that we should be able to hear speech and music from all parts of the earth, not to mention television, he would have been regarded as a madman. And yet to-day these inventions are accomplished facts which we accept as being perfectly natural. Such is the power of science that it invariably surpasses the most daring speculations.

I appeal to these research workers to devise, as I shall attempt to do myself, a mechanical eye, an objective, in a word an apparatus which is lacking, to detect the unknown radiations discussed in this work.

What are we able to perceive with our sense of sight in the immense gamut of radiations ? Nothing but a small zone extending from 375 to 700 trillion vibrations per second. And yet what a social upheaval lies in store for us pending the discovery of this apparatus susceptible of detecting the complete range of waves, known and unknown, which escape our control.

In speaking of man, Descartes said : " I think, therefore I am." This somewhat laconic dictum should not blind us to the fact that man, although superior to animals in many respects, notably in the power of thought, is nevertheless inferior to them, for the time being, in regard to the narrow range of vibrations that he is able to detect. Indeed, man can only see and hear within a very restricted range, and he can only transmit his thought by means of speech. On the other hand, certain animals can travel in a straight line towards a far distant goal, invisible to us, thanks to the vibrations they detect and that our senses cannot perceive.

One of the ways we have of exploring the external world is by means of our visual sense. The eye is the physiological objective which has been admirably copied and which has revealed to us the infinitely small and the infinitely great.

Thanks to a very small gamut in the scale of luminous radiations, we are able to discern the most delicate shades of colours. It is actually the wavelength of each of these colours, of each of these notes of this visual harmony, which excites the cells of our brain, and by the play of multitudinous oscillations, makes them vibrate in unison. So, too, the appearance of certain human beings evokes our sympathy, our love or our contempt. May not these diverse feelings be caused by certain variations in the radiations emitted by these persons ?

This biological eye, admirable creation, has been physically copied and turned into an instrument which captures the luminous rays so as to reproduce through photographs and films, all the sensations experienced directly by the human eye.

Thus for many centuries our unaided vision revealed to us but a small domain of Nature. Man once believed that apart from light and darkness there was nothing to be perceived. In the course of time he became aware of the immensity of the scale of radiations : invisible chemical rays, electromagnetic waves, X-rays, radium emanations and cosmic rays which may still prove to be the most important of all to future researchers. And more particularly, man possessed no sense that could apprehend electric waves, and this realm would have remained for ever closed to him if scientists of genius had not brought into being an " electric eye " which revealed a new world to us all, the world of wireless.

And now, what significance do we discern in the stream of life and in cellular oscillations, and who will invent that eye, that detector of vital oscillations ? When this comes to pass, we shall achieve the mastery of these oscillations. Not only from a biological point of view will these radiations enable us to obtain results of great value to mankind, but also from a social point of view, their practical application may bring about changes of great significance. We shall utilise them for our needs and we shall achieve the transmission of thought and communication with the blind ; we shall know what other people think and we shall communicate with one another, and possibly with animals too, by means of our own radiations. We shall also be able to trace the whereabouts of criminals by the wavelengths of their radiations.

And, indeed, we live in the midst of a mystery, for do we not see birds, insects, and animals of all kinds, devoid of the faculty of speech, yet manifesting powers as marvellous as they are inexplicable ? May we not postulate the existence of thought transmission among all sentient beings ? The instinct of self-preservation in animals is but a verbal expression concealing a reality which is the primary cause of their existence : the whole gamut of radiations, imperceptible to us, is apprehended on their plane for they are capable of emitting and receiving them.

Let us wait hopefully for the day when this superlative eye, this wonderful apparatus that we dream of, will finally appear and reveal in all its complexity and majestic grandeur a new world that science has begun to unveil.[1]

[1] These predictions made by Lakhovsky over a decade ago bear a singular resemblance to the recent speculations on the future of science by one of the most talented members of the Royal Society, Professor J. D. Bernal. In his fascinating work, " The Social Function of Science " (Routledge, London, 1939), Professor Bernal states that language and writing may be superseded by the application of knowledge of the electro-neurology of the brain. People may be able to communicate with one another through an electrical apparatus activated by electrical currents produced in the brain by thinking. (Translator.)

APPENDIX

I. Note on Radium

PAGE

1. Preliminary Remarks 165
2. Professional Criticisms of Radium Therapy . . 167
3. The Tragic Experience of a London Surgeon . 170
4. Symposium of Medical Views on Final Results of Radium Treatment 171
5. Conclusions 174

II. The Multiple Wave Oscillator

Description of Apparatus 178
Photograph of Apparatus 192
Special Note on the Pope's Recovery . . . 181

III. Clinical Reports

Selection of Cases treated with Multiple Wave Oscillator 181
(*a*) Cancer 182
(*b*) Exophthalmic Goitre 190
(*c*) Enlarged Prostate 192
(*d*) Gastro-duodenal Ulcers and other Conditions 195

IV. Effects of Oscillating Circuits on Animals

(*a*) Horses 196
(*b*) Dogs and Cats 196

I. NOTE ON RADIUM

1. Preliminary Remarks

To the public radium is a magic word, to the Press it has news value, and to the Government it appears to be a powerful weapon indispensable for the fight against cancer.

The Minister of Health has recently been empowered by Parliament to " lend money to the National Radium Trust up to a maximum of £500,000 to enable it to purchase radium and other radio-active substances and apparatus required for radio-therapeutic treatment." [1] This sum of public money should at least ensure the material welfare of certain

[1] *The Times* report, December 13th, 1938.

company directors for some appreciable time though it may not prove adequate to provide sufficient radium for the cancer campaign.

Among those who are heretical enough to deny the divinity of monopolies it is well known that the radium traffic is hardly conducted on principles inspired by humanitarian feelings. Contrary to the universal rule in medical practice of withholding no information concerning remedies, the process of extracting radium is kept a secret by its exploiters.

A few years ago an attempt to investigate the radium mystery was made by Dr. Haden Guest, M.P., on behalf of a leading London newspaper. When Dr. Haden Guest arrived at the Chinkolobwe mine in the Belgian Congo he was not permitted to visit the galleries, and the manager would not answer a single question. According to its published statements the Union Minière had made 210 grammes of radium between 1923 and 1929 inclusive, which had been sold at varying prices, working out at an average of £10,000 per gramme, thus amounting to £2,100,000 for seven years' work—a fairly profitable enterprise.

Dr. Haden Guest pointed out that the expenses of the Chinkolobwe mine with its 150 labourers costing about £1 a week and its four or five European officials could not represent a very large part of this sum. He concluded that the public had a right to demand that the Union Minière should abandon its policy of secrecy and publish the figures of the costs of production.

The fantastic price of radium throws an interesting sidelight on the machinations of financial speculators. According to Dr. Haden Guest, shortly after its discovery by Madame Curie, radium could be bought in small quantities at a price of £500 per gramme. In 1910, small quantities of radium were produced from the ore deposits at Joachimsthal in Bohemia, and sold at prices running up to £36,000 per gramme. Until the discovery of radium in Canada in 1932, practically all the world's supplies came from the Belgian Congo and the price was kept up at a very high level, in the region of £15,000 a gramme, but when the Canadian mines began their operations the price fell immediately, sinking as low as £4,000 per gramme. According to the latest reports the National Radium Trust has secured an option to buy radium for the next five years at £4,500 a gramme, excluding containers.

At the present time there are no indications that the radium traffickers will suffer an exposure of their contemptible racket by administrators with a modicum of ethical vision and economic common sense. But there are indications, however, that in the near future radium, as a means of curing cancer, will be ultimately replaced by other methods of treatment, more effective, less dangerous and infinitely less costly than those associated with radium therapy. Meanwhile it should be clearly realised that radium in the treatment of cancer has definite limitations and that its extensive use, even if it were possible, would inevitably result in raising false hopes and in causing much needless suffering.

2. Professional Criticisms of Radium Therapy

The use of radium in the treatment of cancer has become an established part of orthodox medical practice. It is a double-edged weapon capable of destroying cancer cells and also of damaging normal cells. The effects of radium on the human organism are not well understood. Its handling requires special precautions; both patients and medical personnel must be carefully protected at all times or else severe injuries may result, sometimes long after treatment.

It cannot be sufficiently stressed that radium is not, as some enthusiasts would seem to believe, the remedy for cancer "par excellence."

In a recent work on *Radium and Cancer*, H. S. Souttar, surgeon at the London Hospital, states: "*Those who have studied the subject most would be the first to admit that we are still working entirely in the dark, that we really know nothing of the mechanism by which radium produces its effects and that the success we have achieved is the purely empirical result of almost blind trial and error. Moreover, after years of labour and of the most painstaking research, there is no general agreement on the best methods of application or even as to the period over which the treatment should extend. Paris regards as inadequate any period under fourteen days; Stock-*

holm is satisfied that four hours is ample, whilst America, as might be expected, uses a stop-watch." [1]

Souttar also emphasises that "*it cannot be too widely recognised that the use of radium is attended by very definite dangers.*" While admitting that these dangers are usually avoided by taking every possible precaution, Souttar remarks that "*even the most experienced operator will occasionally be surprised by excessive sensitiveness of the patient's tissues and by a reaction altogether out of proportion to that which experience has taught him to expect. . . . Delayed radium necrosis may appear several years after treatment. . . .*"

The disadvantages of irradiation treatment are frankly admitted by one of its leading exponents, Dr. Geoffrey Keynes. He points out "*the possibility of there being a residual tumour after treatment by radium alone* (*of cancer of the breast*) *and the difficulty of knowing whether this contains active cancer or not. . . . In addition to this there is the post-irradiation fibrosis which is apt to appear as long as two years after treatment in the positions where the irradiation has been most intense. Another disadvantage of the conservative method is the increased liability to neuralgia or 'rheumatic' pains in the treated area . . . and it must be remembered that radium needles are dangerous weapons if used with insufficient skill.*" [2]

Further evidence of the dangers of irradiation treatment was recently given in an analysis of 259 cases of radiodermatitis seen in the period 1930–34 inclusive at one of the most famous medical institutions in the world, the Mayo Clinic, Rochester, U.S.A. The report states that "14 per cent. of the patients received their injuries while undergoing radiotherapy for cancer while approximately 10 per cent. of the patients developed cancer at the site of the X-ray or radium dermatitis." The report goes on, stating "This percentage would undoubtedly have been higher if a group of patients whose injuries were of longer duration had been taken." [3]

[1] "Radium and Cancer," H. S. Souttar. London, 1934.

[2] *Brit. Med. Jour.*, October 2nd, 1937.

[3] "Chronic Roentgen and Radium Dermatitis. An Analysis of 259 Cases." T. S. Saunders, M.D., and Hamilton Montgomery, M.D. (*Journal of American Medical Association*, January 1st, 1938, vol. 110, p. 23).

Among physicians, scepticism as to the value of radium in cancer is not uncommon though it may not receive any publicity. In a recent monograph on cancer Dr. Mitchell Stevens, Fellow of University College, London, and consulting physician of the Cardiff Royal Infirmary, condemns irradiation treatment in an emphatic manner. He states: "*I do not think that irradiation treatment is rational, and believe that in the long run it may do more harm than good. Local treatment by surgery, radium and X-rays, even when the local condition is accessible, is only attacking the 'outworks' of the disease, and how long radium and X-rays will be fashionable remains to be seen.*" [1]

In order that medical students, at any rate, may not be misled on the subject of radium therapy in relation to cancer, a standard text-book warns them in the following terms: "*Radium has recently been extensively used in inoperable cases (and also in operable growths) with the modern technique of small doses for prolonged periods, and remarkable results have been reported in that the primary growth has disappeared. It is, however, too early to be sure that the patient's life is really prolonged, as secondary growths will still form, and many recurrences have occurred. We do not consider that radium treatment alone is justifiable in operable growths (of the breast).*" [2]

Finally, another important fact must be stressed concerning the scope of radium in the treatment of cancerous tumours. While radium may have its uses in certain selected cases and under strictly specified conditions, it cannot be successfully employed in cases of cancer of the lungs which is admitted to be increasing, and it is utterly useless in cases of cancer of the stomach [3] which accounts for over 12,000 deaths in this country annually. Hence the necessity of

[1] "Cancer—Its Causation, Prevention and Treatment," W. Mitchell Stevens, M.D. London, 1935.

[2] "The Science and Practice of Surgery," W. H. C. Romanis and P. H. Mitchiner. London. 2 vols., 6th edition, 1937.

[3] Hutchinson, in his "Index of Treatment" (11th edition, 1936), states: "The radical treatment of cancer of the stomach is undoubtedly surgical."

Romanis and Mitchiner concur with this widely accepted rule and add: "In inoperable cases (of cancer of the stomach) any recognised method of treatment such as radium, X-rays, etc., may be considered justifiable, though they are not in the least likely to do any good" ("Science and Practice of Surgery," 6th edition, 1937).

intensive research into the causation of cancer with a view to preventing the occurrence of the disease, for prevention, though an indirect method of attack, is invariably more effective and more scientific, not to say less alarming, than belated attempts at eradicating growing tumours in cancerous patients.

3. The Tragic Experience of a London Surgeon

One of the most significant indictments of radium as a remedy for cancer was recently expressed by the late Dr. Percy Furnivall, consulting surgeon to the London Hospital. In the beginning of 1937 Dr. Furnivall discovered that he had a small tumour on the left tonsillar lingual fold which he diagnosed as cancer. This was duly confirmed by microscopical examination. He immediately consulted an expert and received X-ray treatment supplemented by radium therapy, radon seeds being implanted. Describing his personal experience, Dr. Furnivall wrote : " *I doubt if the results of the modern treatment of malignant disease by the combined methods of X-rays, radium and the surgeon's knife are as well known as they should be.*" He then proceeded to give an account of his own case, and concluded by remarking : " *I would not wish my worst enemy the prolonged hell I have been through with radium neuritis and myalgia for over six months. . . . This account of my own case is a plea for a very careful consideration of all the factors before deciding which is the most suitable method of treatment. . . . The resistance of tissues to radium emanations is still a doubtful quantity, and the intensive research on this subject which is being carried out should be generously supported. This, to my mind, is even more important than buying more radium and using it in cases for which it may not be the most suitable form of treatment.*" [1]

Judging by the professional comments in the medical Press that appeared after Dr. Furnivall's account of his own case, it became clear that there was no generally accepted method of applying radium, that selection of the best method of treatment for each individual case depended

[1] *Brit. Med. Jour.*, February 26th, 1938.

purely on the surgeon's predilection and that the miseries of radium necrosis, neuritis and myalgia were still of frequent occurrence. In his final communication Dr. Furnivall wrote: "*I am surprised at the number of letters I have received from strangers about radium treatment following the publication of my own case. These letters show that disastrous results occur more frequently than I had any idea of, and that patients are not told beforehand of the possibility of such results. I am now more than ever convinced of the necessity of further research on the effects of radium emanations, and care in selecting cases for radium treatment; also the inadvisability of handing out radium for indiscriminate treatment.*" [1]

Soon after writing this Dr. Furnivall died, the victim of cancer, deeply lamented by his colleagues. In the words of the writer of an obituary notice: "He only wished that the members of the profession he had served so well would be careful in the use of a remedy powerful in benefit, but equally powerful to make the remedy worse than the disease."

4. Symposium of Medical Views on Final Results of Radium Treatment

The tragic account of Dr. Furnivall's case evoked many communications from practitioners in all parts of the country. From the North of England a medical man wrote: "*I consider that Dr. Furnivall has done good service in drawing attention to the untoward results of radium and X-ray treatment of malignant disease. . . . For three years all my cases of faucial cancer were sent for radiation treatment by recognised experts. The results were distressing. Though in some there was initial notable improvement in the malignant condition, all relapsed, all suffered severely with pain—some intensely, requiring large and frequent doses of morphine—all had necrosis of soft and bony tissue, and all died within a year, except two. Of these two, one patient, a very early case, died within fifteen months from the date of starting the radiation treatment, with extensive necrosis and with a large fistula; the other died at the end of a little over two years from the time when first seen, with widespread ulceration showing no tendency*

[1] *Brit. Med. Jour.*, March 12th, 1938.

to heal, and with necrosis of about a third of the inner surface of one side of the lower jaw." [1]

A long polemical correspondence in the columns of the *British Medical Journal* followed as a result of Dr. Furnivall's strictures on radium therapy. Undaunted by the lamentable experience of this eminent surgeon, some stalwarts in the radium citadel found little difficulty in vindicating their position by facile arguments and rationalisations, thus asserting their will to believe what they wished to believe, an affliction as old as sin and as incurable. One of these incorrigible doctrinaires with a touch of omniscience confidently predicted that " his (Dr. Furnivall's) ultimate and complete recovery is assured."

A more chastened tone prevailed in the judicial summing-up by a leading Scottish authority on cancer, who observed : " *The original account of Dr. Furnivall served to show from his own experience that the results of radiotherapeutic treatment, like those of other forms of treatment, cannot be stated simply as success or failure according to the survival or death of the individual treated. By his own case-history he demonstrated that the elimination of cancer from certain situations by radiotherapeutic treatment may be associated with such a degree of persistent pain that survival may scarcely be desirable. Surgery likewise, although successful in the sense that the disease has been eradicated, may also demand a heavy price, in disablement, from the survivor. Few radiotherapists of experience would seek to deny that considerable room for improvement exists both in technique and in the assessment of cases in their comparatively young branch of medical science.*" [2]

At this juncture a team of surgeons entered the forum and boldly proclaimed that the probing knife was more potent than the radium needle. These disputants, possibly feeling that the prestige of surgery had suffered a partial eclipse in the warfare against cancer, seized this unparalleled opportunity to launch an attack against the so-called " modern method " of treating cancer known as irradiation. Although certain subjective demands not altogether free from the taint of self-interest may have motivated

[1] *Brit. Med. Jour.*, April 2nd, 1938.
[2] *Brit. Med. Jour.*, April 16th, 1938.

their onslaughts, it must be admitted that the surgeons established their claims with forceful directness. What could be more effective than this marshalling of ineluctable facts by a London surgeon: "*The treatment of cancer by irradiation is receiving an ever-increasing amount of notice in the medical press. Constant references to the subject are also occurring in the lay Press, and the cry is always for radium and yet more radium. The general public is given the impression that all that is necessary to combat the ravages of malignant diseases is a sufficient supply of radium. Brilliant results which have been obtained by other methods are apt to be forgotten and surgery has been driven into the background. Is this fair to the surgeon, or, what is more important, to the patient?*"

Drawing attention to the annual reports of the National Radium Trust and Radium Commission for 1936–1937, whose statistics reveal retrogression rather than advance so far as immunity from recurrence of cancer is concerned, when compared with radical surgery, the writer continues: "*Surgery, in all but inoperable cases, should be the main weapon of attack. The pendulum has swung too far in the direction of irradiation.*" And very pertinently, he concludes: "*It may be said of surgery that no new disease is introduced. Radium, however, may cause œdema, burns, necrosis of soft and bony structures, even osteomyelitis.*" [1]

This variety of disasters caused by radium may well make upholders of irradiation pause awhile, and gullible laymen realise that behind the glamour of radium there are wounds that cannot heal and bones that are dead.

It is also sad to reflect that one of the greatest women of our age, Madame Curie, died of aplastic anæmia due to the effects of radium.

In this question of treatment of malignant disease, eradication of the local tumour is clearly not the only objective. Exponents of various methods of treatment appear to have ignored the axiom that the patient is more important than the disease, and, therefore, as another London surgeon pointed out, our aim should be the cure of the patient with the least possible disturbance to his future well-being. In that respect the application of radium in the treatment

[1] *Brit. Med. Jour.*, June 4th, 1938.

of cancer cannot be said to be immune from certain disastrous consequences inherent in the nature of the substance itself.

5. Conclusions

The foregoing facts have made it abundantly clear that the use of radium, in spite of all possible precautions, is attended by serious risks in a considerable number of cases which cannot be identified beforehand. In view of the controversy on the value of radium that occurred in the medical press recently as the result of a surgeon's tragic experience, the whole position of radium in the treatment of cancer calls for a critical enquiry devoid of any bias or personal predilection.

The fact that the Government has deemed it advisable to give a sum of £500,000 for the purchase of radium makes a dispassionate enquiry still more imperative for, in the long run, such an expenditure may prove to be not only a waste of public money but what is far worse, a sacrifice of human lives.

The necessity of a well-planned strategy in the conduct of the cancer campaign has been urged by a host of medical men in the past, and when at last an effort is made to apply empirical knowledge on an extensive scale, it is important that those in authority should not mistake hypothetical postulates for scientific revelations.

Experienced surgeons of high standing with unimpeachable records of operative cures of many years' duration have condemned irradiation treatment outright and have insisted that surgery should be the main method of eliminating cancer. Assuredly the treatment of choice must be determined by a consideration of final results and not by the dictates of a dominant methodology prone to error and doomed to extinction.

With regard to final results of treatment, the only sound criterion is the survival-rate estimated in periods of five and ten years. Thus, in a series of 172 cases of cancer of the breast of all grades, only 33 cases (or 19 per cent.), survived operation by ten years.[1] This attenuated survivorship is

[1] *Lancet*, September 3rd, 1938.

mainly due to the fact that the great majority of these cases came for treatment at a late stage of the disease, when prospects of cure had been wrecked by procrastination.

In the absence of a non-injurious specific remedy for cancer the claims of surgery cannot be lightly set aside in favour of radium, whose effects on living cells are incalculable in any given individual. Let it be understood, however, that surgical operations for cancer do not constitute an ideal method of treatment, for the shock and mutilation they entail may have far-reaching effects which are apt to be minimised by advocates of the scalpel. But, unlike radium, surgery does not invest itself with a halo of dubious glamour. It performs its function with great skill and achieves its results with commendable rapidity.

In this connexion it may be said that the Press has played a neutral rôle in publishing the facts concerning radium as far as they can be communicated to the public by lay writers. It is hardly an editor's duty to guide public opinion in matters relating to the treatment of cancer and yet sound advice on this tragically urgent question must be given through channels easily accessible to the people. It is deplorable that there should be no permanent liaison officials between the medical profession and the Press to impart knowledge in a spirit of probity and scientific accuracy. The appointment of such experts should be seriously considered by the organisers of the new National Cancer Service which should aim at enlightening the public in a systematic manner as well as providing adequate facilities for treatment.

It has already been pointed out that radium cannot be used for all forms of cancer ; indeed, it is worthless for one of the most prevalent manifestations of the disease, namely, cancer of the stomach. It is, therefore, of the utmost importance that the public should not be given the impression that the solution of the cancer problem is essentially a matter of purchasing a sufficient supply of radium.

The dangers of radium in the treatment of cancer have been repeatedly indicated by many responsible workers, and it is clear that the scope of radium has definite limitations which necessitate the use of other forms of treatment.

Moreover, there are suspicions amounting to convictions, that the exploitation of radium provides a unique opportunity for gross profiteering, a revolting example of human greed which may soon be faced by a well-merited nemesis. Indeed, there are signs that the artificial production of radio-activity such as Neutron rays may put an end to the spectacular evolution of radium both as a weapon against cancer and as an abominable "ramp." Already the National Cancer Advisory Council in Washington have announced that good progress had been made in the experimental treatment of cancer with Neutron rays. The apparatus generating these rays is known as the cyclotron and is actually in use in America and in Denmark. The fact that it is not yet used for treating cancer in London is symptomatic of the lethargy afflicting those whose official duty is to display mental alertness.[1]

The incessant advance of physical science will gradually give us a deeper understanding of the structure of matter, and, incidentally, of the living cell, so that unimagined developments in the treatment of disease will take place. Among such future events, none would be more welcome than the discovery of a substitute for radium, and it may be prophesied that when this comes to pass radium will be swiftly removed from its position of undesirable pre-eminence and consigned to the museum of medical relics like a charm that has lost its power.

M. C.

II. THE MULTIPLE WAVE OSCILLATOR

In his capacity of engineer-physicist Lakhovsky has always taken a keen interest in the construction of electrical appliances. In 1923 he brought out his Radio-cellulo-oscillator with which readers of this work are already

[1] At the annual meeting of the British Empire Cancer Campaign it was stated that a cyclotron had been installed at Cambridge and another at Liverpool, but "it was of urgent importance that at least one should be available in London so that clinical trial could eventually be made of the therapeutic value of neutron rays" (*Brit. Med. Jour.*, December 10th, 1938).

familiar. Subsequently the success of short-wave therapy, like all new therapeutic methods, led to a vogue in which enthusiasm soon outstripped judgment. Contra-indications were gradually discovered, and unexpected effects, injurious to the patient, warned experimenters that the fundamental rule of any medical treatment, *Primum non nocere*, could not be ignored. Indeed, it is now recognised that the indiscriminate application of short waves is not devoid of risks which may have serious consequences. Lakhovsky was prompt to realise the possibility of harmful effects lurking in the midst of encouraging results. He then decided to give up using ultra-short waves capable of causing thermal effects. It occurred to him that better results might be obtained by giving an oscillatory shock to all the cells of the body simultaneously. Such a very brief shock, produced by damped electrostatic waves, does not cause a prolonged thermal effect and therefore cannot injure the cells. Lakhovsky's aim was to produce an oscillatory shock that would cause the diseased cells to oscillate aperiodically, that is to say, not at a specific rate. At first sight, from a physical point of view, the problem seemed insoluble since the human body is made up of something like 200 quintillion cells, each oscillating at a specific rate, typical of its own cellular wavelength. Theoretically this implies the necessity of taking into account as many different wavelengths as there are cells in order that each cell may oscillate in accordance with its own physico-chemical constants.

After many experiments Lakhovsky succeeded in constructing an apparatus generating an electrostatic field in which all frequencies, from 3 metres to the infra-red region, could be produced. Hence, in this field, every cell could find its own frequency and vibrate in resonance. Moreover, it is known that a circuit supplied by damped high frequency currents gives rise to numerous harmonics. These considerations led Lakhovsky to invent an oscillator of multiple wavelengths in the field of which every cell, every organ, every nerve, every tissue, could find its own frequency. To this end he devised a diffuser consisting of a series of separated concentric oscillating circuits connected with one another by silk threads. Thus a type of oscillator was obtained giving all fundamental wavelengths from 10 cm. to 400 metres, corresponding to frequencies of 750,000 to three milliards per second. In addition to this, each circuit emits numerous harmonics which, together with their fundamental waves, interference waves and effluvia, may extend as far as the infra-red and visible light regions (1–300 trillion vibrations per second).

As all cells and even their mitochondria are believed to oscillate within that range of frequencies, they are thus provided, in the field of such an oscillator, with the characteristic individual frequencies enabling them to vibrate in resonance.

In 1931 Lakhovsky brought out his first Multiple Wave Oscillator representing a greatly improved type of his former apparatus, the Radio-cellulo-oscillator, with which geraniums, bearing cancerous tumours, were successfully treated. From 1931 onwards Lakhovsky's new Multiple Wave Oscillator has been used in various Paris hospitals, notably Hôpital Saint Louis, Val-de-Grâce, Calvaire, Hôpital Necker, Dispensaire Franco-Britannique, etc.

The Multiple Wave Oscillator has also been used in most European countries and in America for the treatment of various organic diseases, including cancer. Since its inception in 1931, the Multiple Wave Oscillator has been applied by many workers and no contra-indications nor any harmful effects on patients or medical personnel have ever been reported. This is in striking contrast with short-wave therapy in general, X-rays and radium, whose application, particularly in the case of the latter, has not infrequently been followed by the most serious consequences. Everywhere the results appear to have been most gratifying. A selection of cases treated with this apparatus, together with photographs, is given in the third part of the Appendix.

Photographs of the latest model of the Multiple Wave Oscillator appear on pages 192 and 193.

The apparatus consists of a transmitter and a receiving resonator, both arranged so as to set up an electromagnetic field in their immediate vicinity.

The patient is placed between the two oscillators separated

from each other by a distance of about 4 or 5 feet. The current is then switched on and the apparatus functions instantly.

The duration of treatment and number of applications depend on the state of the patient and the nature of the disease. Generally speaking, a quarter of an hour is sufficient for each application. Excellent results have been obtained by giving a séance of five to seven minutes every other day, but some practitioners advise a longer application, from ten to fifteen minutes.

It should be particularly noted that, unlike the average type of short-wave generator in use in medical practice, the Multiple Wave Oscillator cannot cause any injurious effects. As all the radiations generated by this apparatus are of an electrostatic nature, they cannot overheat or burn the tissues.

The action of the Multiple Wave Oscillator is purely electrical.

SPECIAL NOTE ON THE POPE'S RECOVERY

Information has recently reached this country, and has since been confirmed, to the effect that the miraculous recovery of the Pope in January, 1937, was due to Lakhovsky's Multiple Wave Oscillator, which was applied as a last resort. The treatment was continued for several weeks, and His Holiness was finally able to resume his important duties.

Moreover, the Multiple Wave Oscillator has been in use during the last five years in the Hospital of Vatican City (Bambino Jesu) where many patients have been successfully treated with it.

III. CLINICAL REPORTS

Selection of Cases Treated with Multiple Wave Oscillator

(*a*) *Cancer.*
(*b*) *Exophthalmic Goitre.*
(*c*) *Enlarged Prostate.*
(*d*) *Gastro-duodenal Ulcers and Other Affections.*

At the outset it is important to note that all the following cases were treated with the Multiple Wave Oscillator several years ago, some of them as far back as 1931 when the apparatus was first introduced by Lakhovsky to the medical profession. Without exception, all the patients whose clinical history is given here have remained well, and no recurrence of the original affection has been reported. In the case of cancer this is of such great significance that it is worthy of the most serious consideration.

(*a*) One of the first cases to be treated with the Multiple Wave Oscillator was a case of cancer of the face.

I. *Case of Madame C.*

Age, 68.

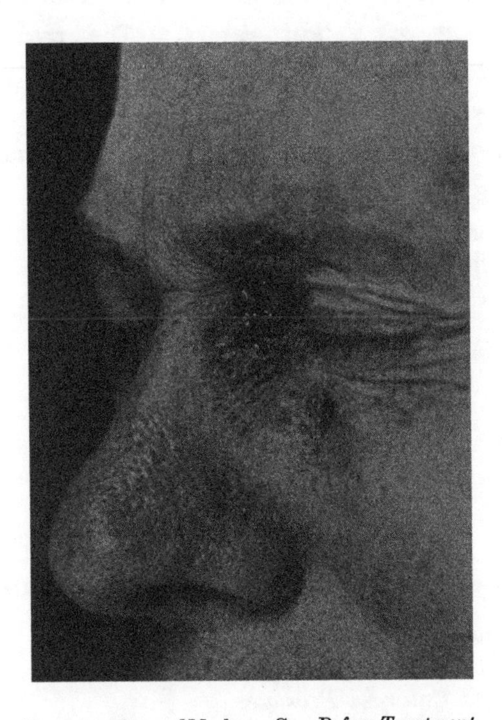

FIG. 1. Case of Madame C.—*Before Treatment.*

Rodent ulcer in inner angle of left eye and sub-orbital region.

Diagnosis. Rodent ulcer (Basal-celled carcinoma) situated in inner angle of left eye, affecting root of nose ;
Diameter, about ½ inch.
Duration, three years.
The diagnosis was confirmed by biopsy (microscopic examination.)

This patient was treated for a facial lesion twenty-three years previously with X-rays. An improvement resulted, but subsequently a suspicious crust developed in the site mentioned above. Treatment with Lakhovsky's Multiple Wave Oscillator began on September 8th, 1931, at the Hôpital Saint Louis. After the third séance, lasting fifteen minutes, there was an improvement in the general state of the patient and a diminution in the size of the lesion.

On November 19th, 1931, the cancerous ulcer had completely disappeared. There was only a scar left without any trace of

FIG. 2. Case of Madame C.—*After treatment.*

induration. The general appearance of the patient showed a remarkable improvement. She stated that she felt rejuvenated and that she had not enjoyed such good health for thirty years.

(Figs. 1 and 2.)

II. *Case of Mr. M. M.*

Age, 80.

Diagnosis. Nævo-carcinoma on left arm.
Axillary glands enlarged.
Diagnosis confirmed by biopsy.
Duration, seven years. No previous treatment.

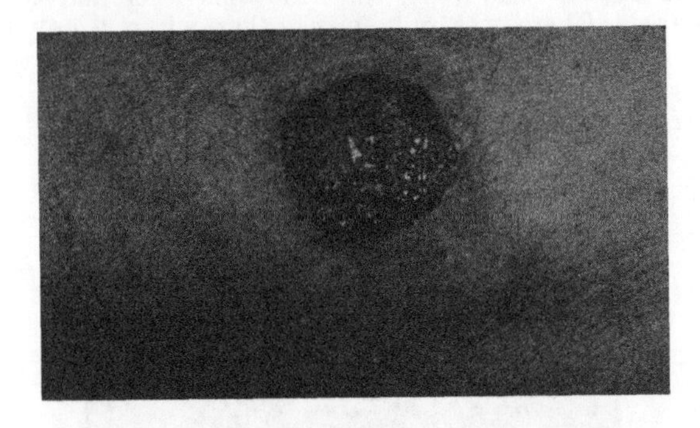

FIG. 3. Case of Mr. M. M.—*Before treatment.* Naevo-carcinoma of left arm.

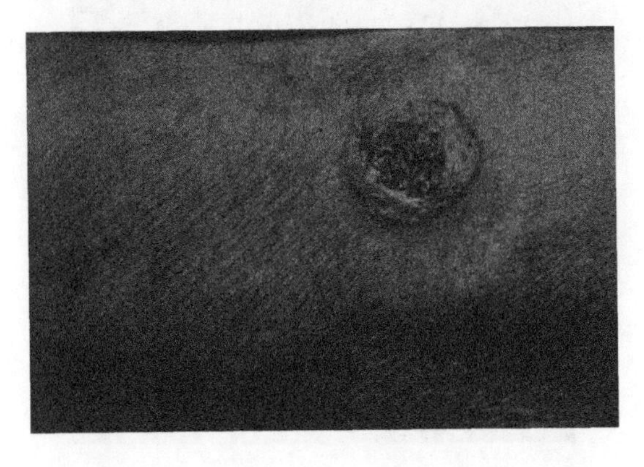

FIG. 4. Same case—After seven applications of the Multiple Wave Oscillator.

Treatment with Multiple Wave Oscillator began on October 9th, 1931, at the Hôpital Saint Louis.

After seven séances, the ulceration was reduced by half.

Six months after commencement of treatment, the lesion had completely disappeared, leaving a clean scar.

(Figs. 3, 4, 5.)

III. *Case of Madame S.* This case is the most striking example of a cancerous tumour cured by means of Lakhovsky's Multiple Wave Oscillator, after operation and radium had both failed to eradicate the disease. When it is realised that this case was cured within a few weeks, the result becomes still more significant.

This old lady, aged 82, had been treated three years previously in an anti-cancerous centre. After an operation performed there in 1929, an ulcerated lesion of neoplastic nature developed. In the course of 1929 and 1930, radium was applied. A temporary improvement followed, but the tumour persisted and began to grow rapidly. Another small tumour was also observed in the sub-orbital region.

FIG. 5. Same case—Three months after treatment. After six months' treatment this highly malignant tumour had completely disappeared leaving a smooth scar.

Diagnosis. Epithelioma of upper part of left cheek.
Size, $2\frac{1}{2} \times 1\frac{1}{4}$ inch.

As the general condition of the patient was gradually becoming worse, she was sent to the Calvaire Clinic, known in Paris under the more gruesome name of "ante-chamber of the cemetery."

Treatment with the Multiple Wave Oscillator began on April 26th, 1932, and lasted fifteen minutes. After only two applications an improvement was observed. With further treatment the improvement was maintained, and on May 12th, 1932, a final application of twenty minutes' duration was given. The enlarged sub-maxillary glands and œdema, noticed at the time of examination before treatment began, were no longer present.

The patient was photographed on May 30th, 1932.

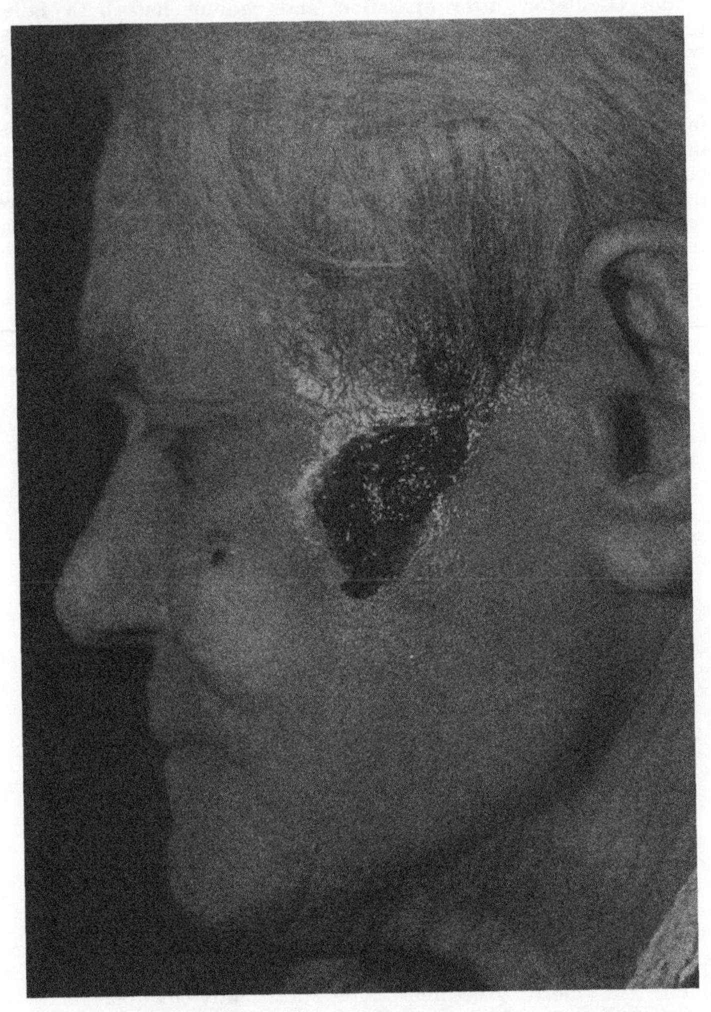

Fig. 6. Case of Madame S.—*Before treatment.* Epithelioma of upper part of left cheek. Radium was applied but no benefit resulted. Photograph taken April 25th, 1932.

It will be observed that not only has the tumour completely disappeared but that the skin shows distinct signs of rejuvenation as evidenced by the smoothed-out wrinkles in the face and neck. (Figs. 6, 7.)

FIG. 7. Case of Madame S.—*After treatment.* Photograph taken May 30th, 1932, showing final result after only three weeks' treatment.

It will be noticed that not only has the cancerous tumour disappeared, but that the skin of this old lady of 82 shows distinct signs of rejuvenation.

IV. *Case of Mr. J. S.*

Age, 61.

Diagnosis. Baso-cellular carcinoma in inner angle of left eye. Diagnosis confirmed by biopsy.

Duration. Fifteen years. No previous treatment.

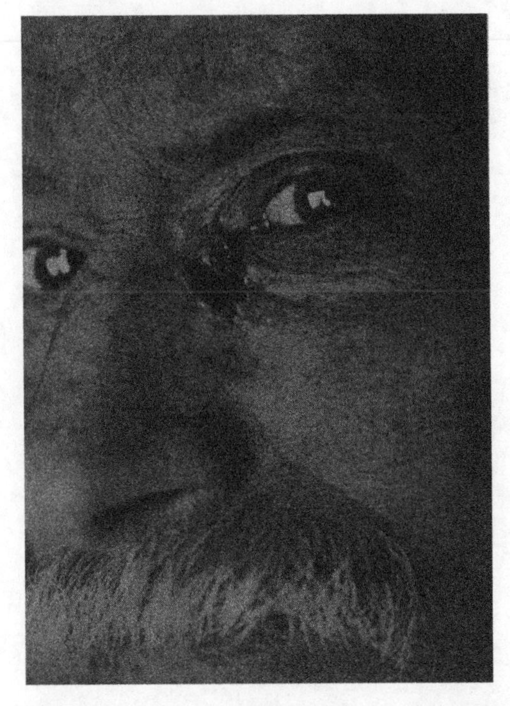

FIG. 8. Case of Mr. J. S.—*Before treatment.* Rodent ulcer in inner angle of left eye.

Treatment with Multiple Wave Oscillator began on October 13th, 1931, at the Hôpital Saint Louis. On December 29th, 1931, the lesion was covered with a scar. Subsequently the patient received further treatment, and in the course of 1932 a complete cure resulted.

The patient stated that he felt greatly rejuvenated and that he could undertake heavy manual labour without experiencing fatigue.

(Figs. 8, 9.)

FIG. 9. Case of Mr. J. S.—*After treatment.*

V. *Case of Radium Burn.*

From a medical point of view this case is of great importance, for it demonstrates both the dangers of radium and the regenerating effects of Lakhovsky's Multiple Wave Oscillator.

The diagnosis of this case was radium dermatitis.

The lesion originated from an ordinary wart on the middle finger. Radium was applied with the result that a severe burn appeared which resisted all forms of treatment for a long period. The tendon was partially necrosed and the patient complained of local pain and rigidity.

After treatment with the Multiple Wave Oscillator for some months the radium burn was healed and mobility of the finger largely restored.

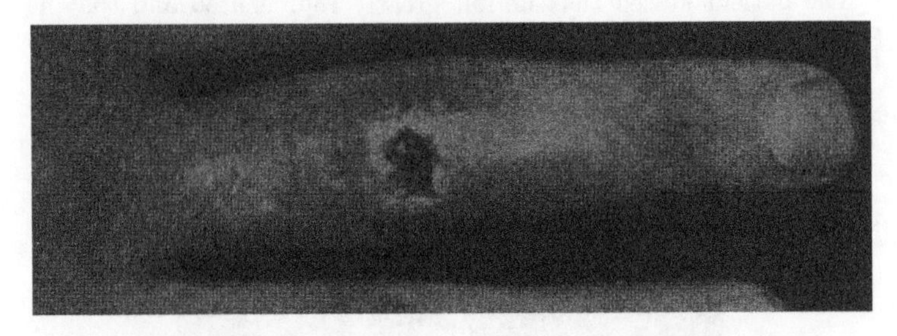

FIG. 10. Case of Radium burn—*Before treatment.*

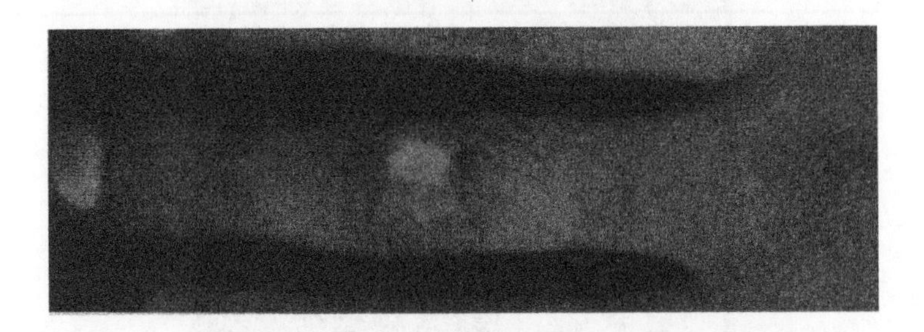

FIG. 11. Case of Radium burn—*After Treatment.* Lakhovsky's Multiple Wave Oscillator succeeded in healing this notoriously incurable lesion after all methods of treatment had failed. (Case of Professor Sven Johansson, Goeteborg, Sweden.)

This case was treated by Professor Sven Johansson in a Hospital Clinic at Goeteborg (Sweden).

(Figs. 10, 11.)

(*b*) *Case of Exophthalmic Goitre.*

The patient, a middle-aged woman, was first examined at the Institut de Physique Biologique in Paris, in January, 1938, where she was treated with the Multiple Wave Oscillator.

Diagnosis. Exophthalmic goitre.

Duration. Eleven years.

Operation having been refused, radio-electrical treatment was decided upon.

The first application was given on January 11th, 1938. After a few applications the general condition of the patient was greatly improved and the size of the goitre considerably reduced.

Treatment with the Multiple Wave Oscillator was continued throughout February. The patient was photographed on March 4th, 1938.

FIG. 12. Case of exophthalmic goitre—*Before treatment.*

FIG. 13. Case of exophthalmic goitre—*After treatment.* Photograph taken on March 4th, 1938. This remarkable result with Lakhovsky's Multiple Wave Oscillator was achieved after only seven weeks' treatment. It will be observed that the goitre has completely disappeared and the general condition of the patient seems to be excellent.

It will be observed that the goitre has completely disappeared and the general condition of the patient seems to be excellent. (Figs. 12, 13.)

(*c*) *Enlarged Prostate.* Hypertrophy or enlargement of the prostate gland is one of the most serious conditions afflicting elderly and aged men. An enlarged prostate invariably necessitates surgical intervention. So far no medical treatment has been found to be curative, and in the vast majority of cases the only hope of saving the patient's life lies in an operation entailing certain risks and complications. In view of the grave prognosis associated with cases

Fig. 14. Lakhovsky's Multiple Wave Oscillator.

of enlarged prostate, reports of cures achieved by means of Lakhovsky's Multiple Wave Oscillator acquire a significance of exceptional import.

The following clinical reports by reputable medical men have established the fact that prostatic enlargement has been cured by the systematic application of Lakhovsky's Multiple Wave Oscillator and the patients restored to health.

Only a selection of cases can be given here. The results, however, are so impressive that this method of treatment should receive immediate attention in this country.

Case I.

Under the care of Professor de Cigna, of Genoa

Patient aged 64.

The diagnosis was made by an eminent specialist. The prostate

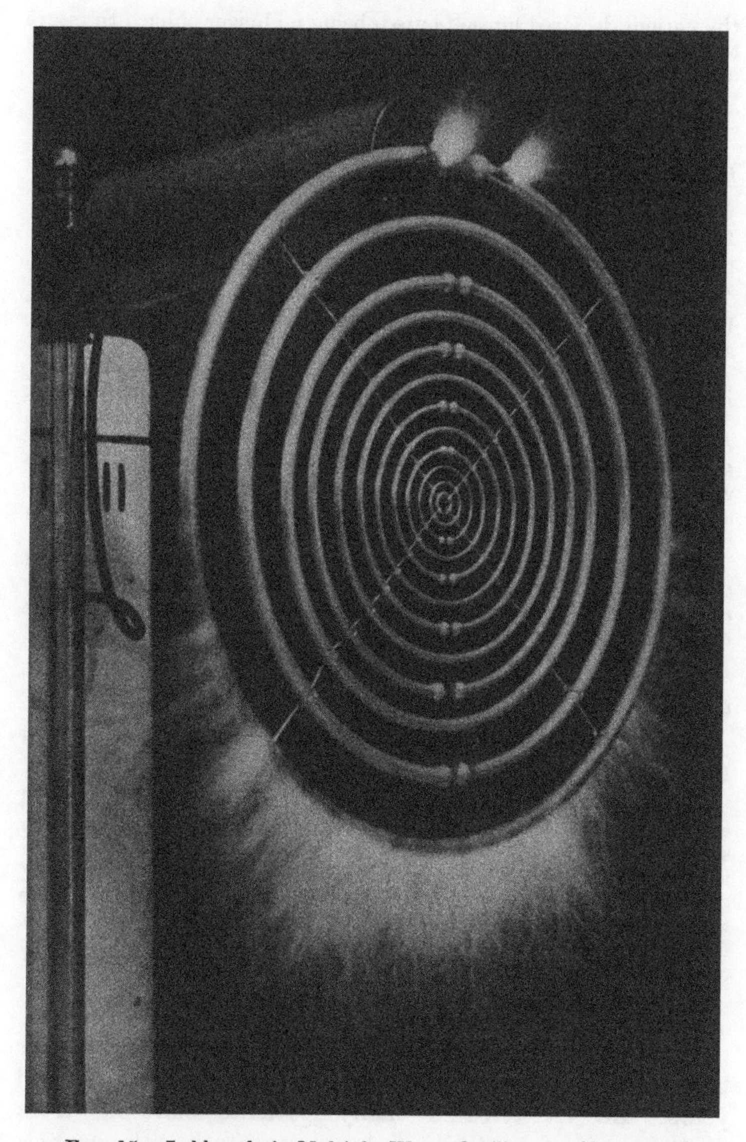

FIG. 15. Lakhovsky's Multiple Wave Oscillator. Close-up of Transmitter in action, showing *effluve* (electric brush).

was found to be the size of a small orange. The patient refused operation and was compelled to lead a "catheter life." After ten applications of the Multiple Wave Oscillator spread over two months, the patient declared himself cured, being no longer obliged to resort to the catheter. Examination by the same specialist six weeks after the last application showed, to his astonishment, that the prostatic enlargement had disappeared.

Case II.

Under the care of Dr. Rigaux, of the Institut de Physique Biologique, Paris.

Patient aged 62.

Diagnosis made by a urologist who advised immediate operation. Prostate greatly enlarged, size of a small orange. Patient had to resort to a catheter.

Treatment with Multiple Wave Oscillator began on July 16th, 1932. Two applications were given daily of five minutes' duration.

The volume of urine began to increase after a few applications. At the end of three weeks' treatment the patient appeared to micturate normally. On re-examination by the urologist the prostate was found to be normal in size. Being somewhat incredulous about the permanence of the result, the urologist suggested another examination six months later.

The improvement was duly maintained, and up to the present time, that is to say six years after the commencement of treatment, there has been no recurrence and the patient is in excellent health.

Case III.

Under the care of Dr. Rigaux, Paris.

Patient aged 60.

Diagnosis was made by three eminent specialists who all advised an immediate operation. Catheterisation had been necessary for some time. The patient decided to give the Multiple Wave Oscillator a trial. Treatment began in October, 1934, in séances of ten minutes' duration. At the end of three weeks the patient was able to resume his occupation and the improvement was maintained. For the past four years there has been no recurrence and the patient's health continues to be excellent.

Dr. Rigaux's experience with the Multiple Wave Oscillator in cases of enlarged prostate extends over a period of several years. He has treated many patients suffering from disturbances of micturition due to prostatic enlargement, and has noted an improvement in many cases while normal function with marked diminution of the gland has often resulted.

For the past five years another medical practitioner,

Dr. Henry, of Brussels, has treated many cases of enlarged prostate with Lakhovsky's Multiple Wave Oscillator, and has achieved a remarkable measure of success. In some cases, frequency has been considerably alleviated, while in others, the prostate, which on examination was found to be greatly enlarged, was gradually reduced to normal proportions after treatment, and the patients restored to health and enabled to resume their former activities.

(*d*) *Gastro-duodenal Ulcers and other Affections treated with the Multiple Wave Oscillator.* Since 1934 Professor de Cigna of Genoa has treated several hundred patients with Lakhovsky's Multiple Wave Oscillator, and has become a keen advocate of this method of treatment.

Recently he has reported a series of cases to the Royal Academy of Medicine of Genoa which formed the subject of a special communication at the International Short-Wave Congress in Vienna in 1937.

All these patients were subjected to strict clinical examination supplemented by biopsy and radiograms.

The cases treated by Professor de Cigna cover a great variety of different conditions, including basal-celled carcinoma, lupus erythematosus, otitis media, gynæcological affections and prostatic enlargement. A number of cases were photographed before and after treatment, and full accounts duly appeared in the Italian medical Press. Furthermore, functional and psychological conditions, such as asthma, insomnia, neuralgia, etc., were also treated with pronounced success.

In addition to this, Professor de Cigna has drawn attention to the beneficial effects of the Multiple Wave Oscillator in patients suffering from gastro-duodenal ulcers.

The results he has obtained in a series of cases, all radiologically controlled, were so remarkable that they were reported to the Royal Academy of Medicine of Genoa.

The full clinical reports of Professor de Cigna's cases, referred to in these pages, appeared in his communication to the International Short Wave Congress held in Vienna, July 12th–17th, 1937.

IV. EFFECTS OF OSCILLATING CIRCUITS ON ANIMALS

(*a*) *Horses.*

(*b*) *Dogs and Cats.*

(*a*) The following observations, based on experiments carried out at the Institut physiologique de la Croix-Blanche at Vaucresson (France), were reported by P. Fournier-Ormonde.

Having noticed an access of vigour as a result of the application of one of Lakhovsky's oscillating circuits on himself, Fournier-Ormonde, in his capacity of horse-breeder, decided to experiment on horses.

For the first experiment, seven horses were chosen and fitted with oscillating circuits in the form of collars measuring about 30 inches in circumference. In all cases a distinctly invigorating effect on the horses became apparent and a favourable influence on the fertility of certain mares was also observed. The most interesting case was one of rejuvenation of a 24-year-old stallion by means of an oscillating circuit.

Application of the circuits round the horses' necks and pasterns was followed, after a variable period of time depending on the condition of the animals, by unmistakable signs of improvement in their general condition. The eyes became brighter, the coat more glossy, and the skin more supple ; in fact the horses appeared to be " full of beans," as manifested by their spontaneous galloping.

Analysis of the horses' blood after treatment provided more evidence in favour of using oscillating circuits for ailing animals.

(*b*) From a veterinary point of view the application of Lakhovsky's oscillating circuits appears to have been as successful as in the case of human beings. Sick, injured and ageing animals have been reported much improved or cured, and the list of treated animals includes horses, dogs, cats, and even an old parrot.

It would seem, therefore, that this eminently humane form of treatment should commend itself to all animal lovers.

M. C.

INDEX OF NAMES

ANDERSON, 14
Attilj, Sordello, Professor, 143, 145, 146
Aubert, 46

BACON, 20
Baxendall, 116
Behla, 26
Bernal, T. D., Professor, 164
Berthelot, 74, 110
Blackett, Professor, 14
Blandford, 117
Boubée, Nérée, 124
Bowen, 111
Broglie, de, 77

CAMERON, DR., 111
Campbell, 108
Casamajor, M. J., 33
Cathelin, 46
Cigna, de, Professor, 192, 195
Clerk Maxwell, 6, 55
Cope, John, 23
Cremonese, Guido, Professor, 12, 13
Crile, George, Dr., 28, 29
Cumberbatch, E. P., 7
Curie, Madame, 24, 273
Cyon, de, 50

D'ARSONVAL, 99, 119, 126, 143
Darwin, 153
Descartes, 162
Dubois, Raphael, 158

FABRE, 39, 41
Faraday, 6
Faure, Maurice, Dr., 121
Franck, 12, 76
Furnivall, Percy, Dr., 170, 171, 172

GABOR, 12
Galileo, 15
Galvani, 6
Geitel, 108
Gilford, Hastings, 21, 22
Gill, C. A., Colonel, 16
Göckel, 111
Gosset, 84
Guest, Haden, Dr., 166
Gurwitsch, 12, 76
Gutmann, 84

HAECKEL, 153
Hansemann, 153
Hartmann, Dr., 137
Haviland, Alfred, Dr., 26, 27
Heaviside, 109
Helmholtz, 51
Henneguy, 159
Henry, Dr., 195
Herschel, William, Sir, 16, 116, 118
Hertwig, 153
Hertz, 66
His, 153

JEANS, JAMES, SIR, 14, 15, 116
Johansson, Sven, Professor, 190

KEYNES, GEOFFREY, DR., 168
Kolb, 26
Koppen, 117

LAKHOVKSY, 1 *et seq.*; 28, 29; 41, 84 *et seq.*; 88 *et seq.*; 99 *et seq.*; 119 *et seq.*; 113, 132; 126, 141; 176 *et seq.*; 192, 193
Linnæus, 35
Lockyer, Norman, Sir, 16

MAGROU, 84
Marconi, 6
Marie, Auguste, Dr., 96
Masse, 46
Maxwell, Clerk, 6, 55
Meldrum, Dr., 16, 116
Millikan, 13, 111, 112, 132
Mitchiner, 169
Moreux, 117, 118
Morrell, Conyers, Dr., 17
Moynihan, Lord, 20

Naegeli, 153
Nichols, 154
Nodon, Albert, 74, 75, 76

Occhialini, 14
Ormonde, Fournier-, P., 196

Pohl, von, 26

Quinet, Dr., 45

Regener, 14
Reiter, 12
Riccioli, 116
Rigaux, Dr., 194
Roffo, Professor, 90
Romanis, 169
Röntgen, 24
Ross, Ronald, Sir, 23
Russo, M., Dr., 125

Sardou, G., Dr., 121
Sesari, M., 99
Shannon, W. J., Dr., 138
Simeray, Dr., 139
Smith, Erwin F., 84
Souttar, H. S., 167, 168
Spencer, 153
Stélys, M., 125
Stevens, Mitchell, Dr., 169

Tchijevsky, 17
Tear, 154
Ternier, 46
Thomas, Dr., 154

Vincent, M. P., 114
Vries, de, 153

Weismann, 153
Wiesner, 153
Wilson, 108

SUBJECT INDEX

Animals, effects of oscillating circuits on, 196
problem of instinct in, 31 *et seq.*
Antennæ, in insects, 36, 38
Artesian wells, 138
Aurora borealis, 15, 16, 116
Auto-electrification, in living beings, 43 *et seq.*

Bat, effects of radiations on, 34, 35
Biomagnomobile, units, 29, 94, 140, 157, 158, 161
Birds, nocturnal, 34
rôle of orientation in flight of, 45
Bombyx, Oak-, diurnal experiments with, 39, 41
Burying-beetles, activities of, 42
Butterfly, great-peacock-, nocturnal experiments with, 38

Cancer, and hydrocarbons, 91
and Neutron rays, 176
and oscillatory disequilibrium, 24
and soil, 25 *et seq.*
and water, 136 *et seq.*
etiology of, 126
experimental, in plants, 84 *et seq.*
final results of surgical treatment, 174
geological and geographical distribution of, 126 *et seq.*
Lakhovsky's theories on causation of, 24, 25
pathology of, 88 *et seq.*
limitations of orthodox methods of research, 21 *et seq.*
mortality in England and Wales, 19
National Cancer Service, 20, 175
the problem of, 17 *et seq.*
research, criticism of methods, 22, 23
Cancerous diseases, geographical distribution in British Isles, 27
Capacity, electrical, definition, 9
in birds, 44, 45
rôle of, 57
Carrier pigeons, powers of orientation, 32 *et seq.*
Cell, differentiation, and heredity, 153
formation of primordial, 151
phases of indirect division of, 72
Cellular energy, infinitesimal value of oscillating, 154
oscillation, therapeutics of, 140 *et seq.*
radiation, characteristics and wavelengths, 71 *et seq.*
nature of, 74 *et seq.*
Cholesterol, irradiated, and cancer, 91
Circuits, oscillating. *See* Oscillating circuits.
Conducting soil, 133
substances, 160
Conductivity, 107
Cosmic radiation, 13 *et seq.*
in relation to nature of soil, and cancer, 130 *et seq.*
rays, 110 *et seq.*
Cyclotron, 176

Differentiation of cell, and heredity, 153

Electrical capacity, in birds, 44
rôle of, 57
Electrification by friction of wings, 43
Electromagnetic fields, induction of, in cells, 157
waves, phenomena associated with, 56
table of, 56
Electro-Radiobiology, First International Congress, 7
Elements, characteristics of living species, 152
Energy, nature of radiant, 102 *et seq.*

Fever, and its function, 93 *et seq.*

Geographical distribution of cancerous diseases in the British Isles, 27
Geological and geographical distribution of cancer, 126 *et seq.*

Geological map of Paris, 128
nature of soil, and cancer, 27
Geraniums, inoculated with cancer, 104
treated with Radio-cellulo-oscillator, 86 *et seq.*
with oscillating circuits, 104, 144
Globulins, chemical composition of, 90
formation of, 88 *et seq.*
Glow-worm, radiations of, 53

Heredity, and cell differentiation, 153 *et seq.*

Inductance, definition, 8
Induction, in fixed oscillating fields, 155
of electromagnetic fields in cells, 157
self-, *rôle* of, 57
Instinct, general considerations, 31
of orientation, 32
Institutional research, criticisms, 21, 23
Insulating soil, 134
substances, 160
Interference phenomena, 121, 122
Ionisation, 107

Lemmings, activities of, 35
Life, origin of, 149 *et seq.*

Meteorology, as link between physics and biology, 118
Microbes, action of, on living cells, 159, 160
electrical properties of, 79
oscillatory action of, 79
Microbiology, 99
Migration, explanation of, 46
Mitogenetic rays, 12, 76
Moon, influence of, on hospital patients, 147
Multiple Wave Oscillator, 11, 192, 193
and Pope's recovery, 181
and treatment of cancer, 25, 181 *et seq.*
description of apparatus, 177–179
selection of cases treated with,
(*a*) cancer, 181–190
(*b*) enlarged prostate, 192–195
(*c*) exophthalmic goitre, 190–191
(*d*) gastro-duodenal ulcers and other affections, 195

Necrophorus (Burying-beetles), activities of, 42
Neutron rays, 176

Oak-bombyx, experiments with, 39, 41
Orientation, instinct of, 32
rôle of, in flight of birds, 45
Oscillating circuit, comparison of living cells to, 69
constitution of cellular, 70, 150
definition, 8
diagrams, 9, 10
circuits, effects on animals, 196
experiments on human beings, 145 *et seq.*
on plants, 104 *et seq.*
of Hertz, diagram, 66
fields, induction in fixed, 155 *et seq.*
Oscillation, mechanism of cellular, 10, 11
of living cells, influence of astral waves, 119
therapeutics of cellular, 140 *et seq.*
Oscillations, electrical, explanatory analogies, 57 *et seq.*
Oscillator, Multiple Wave. *See* Multiple Wave Oscillator.
Radio-cellulo, 10, 83, 84, 86
Oscillatory action of microbes, 79
disequilibrium and cancer, 24

Peacock-butterfly, experiments with, 38
Pigeons, carrier, powers of orientation, 32 *et seq.*
Pyretotherapy, 95

Radiant energy, nature of, 102
Radiation, cellular, characteristics and wavelengths, 71 *et seq.*
nature of, 74 *et seq.*
cosmic, 13 *et seq.*
nature of soil in relation to, 130 *et seq.*
and physiological effects, 14
mitogenetic, 12, 76
nature of, in living beings, 49 *et seq.*
penetrating, 108, 110
solar, and photolysis, 109
Radiations, nature and characteristics of, 55
of living beings, 11, 12, 49, 102
solar, and remarkable vintage years, 120
war of, 11

Radio-cellulo-oscillator, 10, 83, 84, 86
Radiogen, 29
Radiogoniometer, 37
Radiographs, of living beings, 76
Radium
(*a*) Analysis of 259 cases of radium dermatititis, 168
(*b*) Note on Radium, 165 *et seq.*
(*c*) Prices of Radium, 166, 167
(*d*) Symposium of medical views, 171 *et seq.*
(*e*) Tragic experience of London surgeon, 170, 171
therapy, professional criticisms of, 167 *et seq.*
Rays, cosmic, in relation to life, 110 *et seq.*
penetrating powers of, 14

Self-induction, rôle of, 57
Semi-circular canals, in birds, *rôle* of, 36 *et seq.*
in different species of vertebrates, 51
Soil, and cancer, 25 *et seq.*
conducting, 133
influence of nature of, and cancer, 123 *et seq.*
insulating, 134
nature of, in relation to cosmic radiation and causation of cancer, 130 *et seq.*
Soils, carcinogenic, 125
Spectrum of the living, 29
Sunspots, 15 *et seq.*
and biological phenomena, 118
and cosmic radiation, 114 *et seq.*
and pandemics of malaria, 16
and physical phenomena, 116
and waves of epidemic diseases, 17

Temperature, in human body, significance of, 93 *et seq.*
Toxins, 158

Ultra-violet radiation, in human body, rôle of, 29

Universion, and cosmic rays, 110, 112
definitions, 29, 112

Vintage years, correlated with sunspots, 120, 121

Water, in relation to cancer, rôle of, 136 *et seq.*
sterilisation by metals, 98 *et seq.*
Waves, astral, influence of, 119
cosmic, influence of soil on field of, 123 *et seq.*

PRINTED IN GREAT BRITAIN BY THE WHITEFRIARS PRESS LTD.
LONDON AND TONBRIDGE

S.L. P

CPSIA information can be obtained
at www.ICGtesting.com
Printed in the USA
BVHW032039180919
558788BV00003B/296/P